# Knowledge-Driven Board-Level Functional Fault Diagnosis

Fangming Ye · Zhaobo Zhang
Krishnendu Chakrabarty · Xinli Gu

# Knowledge-Driven Board-Level Functional Fault Diagnosis

 Springer

Fangming Ye
Huawei Technologies
Santa Clara, CA
USA

Zhaobo Zhang
Huawei Technologies
Santa Clara, CA
USA

Krishnendu Chakrabarty
Department of Electrical and Computer
    Engineering
Duke University
Durham, NC
USA

Xinli Gu
Huawei Technologies
Santa Clara, CA
USA

ISBN 978-3-319-82054-5          ISBN 978-3-319-40210-9    (eBook)
DOI 10.1007/978-3-319-40210-9

Printed on acid-free paper

This Springer imprint is published by Springer Nature
The registered company is Springer International Publishing AG Switzerland

*To my beloved family for their boundless support.*

Fangming Ye

*To those days and nights at graduate school.*

Zhaobo Zhang

# Preface

The semiconductor industry continues to relentlessly advance silicon technology scaling into the deep-submicron (DSM) era. High integration levels and structured design methods enable complex systems that can be manufactured in high volume. However, due to increasing integration densities and high operating speeds, subtle manifestation of defects leads to functional failures at the board level. Functional fault diagnosis is, therefore, necessary for board-level product qualification. However, ambiguous diagnosis results can lead to long debug times and wrong repair actions, which significantly increase repair cost and adversely impact yield.

A state-of-the-art diagnosis system involves several key components: (1) design of functional test programs, (2) collection of functional-failure syndromes, (3) building of the diagnosis engine, (4) isolation of root causes, and (5) evaluation of the diagnosis engine. Advances in each of these components can pave the way for a more effective diagnosis system, thus improving diagnosis accuracy and reducing diagnosis time. Machine-learning techniques offer an unprecedented opportunity to develop an automated and adaptive diagnosis system to increase diagnosis accuracy and speed.

This book provides a comprehensive set of characterization, prediction, optimization, evaluation, and evolution techniques for a diagnosis system. Readers with a background in electronics design or system engineering can use this book as a reference to derive insightful knowledge from data analysis and use this knowledge as guidance for designing reasoning-based diagnosis systems. Meanwhile, readers with a background in statistics or data analytics can use this book as a case study for adapting data mining and machine-learning techniques to electronic system design and diagnosis.

This book identifies the key challenges in reasoning-based board-level diagnosis system design, and presents machine-learning-based solutions and corresponding results that have emerged from cutting edge research in this domain. It broadly explores a series of topics ranging from high-accuracy fault isolation, adaptive fault isolation, diagnosis system robustness design, system performance analysis and evaluation, knowledge discovery, and knowledge transfer.

This book first describes a diagnosis system based on support-vector machine (SVM), multi-kernel SVM (MK-SVM), and incremental learning. The MK-SVM method leverages a linear combination of single kernels to achieve accurate root-cause isolation. The MK-SVMs thus generated also can be updated based on incremental learning. Furthermore, a data-fusion technique, namely majority-weighted voting, is used to leverage multiple learning techniques for diagnosis.

The diagnosis time is considerable for complex boards due to the large number of syndromes that must be used to ensure diagnostic accuracy. Syndrome collection and analysis are major bottlenecks in state-of-the-art diagnosis procedures. Therefore, this book describes an adaptive diagnosis method based on decision trees (DT). The number of syndromes required for diagnosis can be significantly reduced compared to the number of syndromes used for system training. Furthermore, an incremental version of DTs is used to facilitate online learning, so as to bridge the knowledge obtained at test-design stage with the knowledge gained during volume production.

This book also includes an evaluation and enhancement framework based on information theory for guiding diagnosis systems using syndrome and root-cause analysis. Syndrome analysis based on subset selection provides a representative set of syndromes. Root-cause analysis measures the discriminative ability of differentiating a given root cause from others. The metrics obtained from the proposed framework can provide guidelines for test redesign to enhance diagnosis. In addition, traditional diagnosis systems fail to provide appropriate repair suggestions when the diagnostic logs are fragmented and some syndromes are not available. The feature of handling missing syndromes based on imputation methods has therefore been added to the diagnosis system.

Finally, to tackle the bottleneck of data acquisition during the initial product ramp-up phase, a knowledge-discovery method and a knowledge-transfer method are proposed for enriching the training data set, thus facilitating board-level functional fault diagnosis.

In summary, this book targets the realization of an automated diagnosis system that offers the benefits of high accuracy, low diagnosis time, self-evaluation, self-learning, and ability of selective learning from other diagnosis systems. Although the goal of this work was to advance board-level diagnosis, the core techniques developed in this book can also be leveraged for electronic systems beyond the board level.

Santa Clara                                                                                          Fangming Ye
Santa Clara                                                                                        Zhaobo Zhang
Durham                                                                            Krishnendu Chakrabarty
Santa Clara                                                                                               Xinli Gu

# Acknowledgments

The authors acknowledge the support received from Huawei Technologies. The authors also appreciate the contribution of Shi Jin to the work on handling missing syndromes, described in Chap. 6.

# Contents

# Chapter 1
# Introduction

Relentless technology scaling has increased the performance and complexity of electronic products by orders of magnitude in the past few decades. A complex system today consists of several chassis, and each of them contains a number of printed circuit boards (PCBs). A typical board consists of many application-specific integrated circuits (ASICs) and memory devices. Each ASIC in turn consists of hundreds of inputs/outputs (I/Os), millions of logic gates, and several tens of millions of bits of embedded memory. Moreover, the operating frequencies of high-speed ASICs are above 1 GHz, and the data rates of high-speed I/Os are up to 6 Gbps [1, 2]. With increasing complexity and higher speeds, defective-parts-per-million (DPPM) rates continue to increase and subtle functional failures are becoming increasingly difficult to detect and diagnose for root-cause identification [3, 4].

In this chapter, we provide motivation for the book and introduce basic concepts and terminology. Section 1.1 presents an overview of manufacturing tests and diagnosis. Section 1.2 discusses the state-of-the-art in board-level diagnosis. Section 1.2.4 presents the challenges of board-level diagnosis and motivation for this book. Finally, an outline of this book is provided in Sect. 1.3.

## 1.1 Introduction to Manufacturing Test

### 1.1.1 System and Tests

This section presents some basic definitions that are used in system test and diagnosis [5, 6].

- **Definition 1**. A *system* S is defined as a heterogeneous collection of cooperating entities, designed to realize a specified group of functionalities.

© Springer International Publishing Switzerland 2017
F. Ye et al., *Knowledge-Driven Board-Level Functional Fault Diagnosis*,
DOI 10.1007/978-3-319-40210-9_1

Definition 1 is quite general and it can be adopted to cover a set of different types of systems in different fields of engineering, from digital systems to computer networks, nuclear facilities, and chemical plants [5, 6].

- **Definition 2**. A *fault* **f** is defined as a source of misbehavior of a system. The presence of a fault puts the system in a state where the execution of one or more of its functionalities is different from the expected execution.

In the reliability literature [7], there is a distinction between *defects*, which correspond to locations of a system instance containing a difference between specification and implementation, *faults*, defined as the actual causes of a system misbehavior, and *failures*, representing the external observable effects of a fault, producing a deviation of functionality from specifications.

- **Definition 3**. *Diagnosis* refers to the process of determining the causes of system failure.

In order to describe the diagnosis process, some additional definitions are needed, covering both the targets of this process (*components*) and the information required to be collected to execute diagnosis (*tests*). These concepts will be introduced later.

- **Definition 4**. A *component* **c** represents a part of system.

Let $\mathbf{C_S} = \{\mathbf{c}_1, \mathbf{c}_2, \mathbf{c}_3 \ldots \mathbf{c}_n\}$ be the set of all components that constitute system **S**. A component containing (at least) one fault is referred to as a *root-cause* component; every other component is *fault-free*.

Definition 4 requires that each component must be identified unambiguously within the system. An extension of Definition 2 is that a fault can affect only one component, and two components of a system cannot overlap.

According to the above definitions, a component can correspond to a physical device in the system (e.g., a memory chip in a digital system). However, this condition is not strictly required. A component can also refer to a group of physically connected elements within the system, or even a set of noninterconnected entities contained in a system (*virtual component*).

- **Definition 5**. A *subcomponent* **sc** represents a part within a component **c** of system **S**.

Note that the relationship between components and subcomponents is similar to the relationship between system and components, since it introduces a partition of the fault set of each component.

- **Definition 6**. A (diagnostic) *test* **t** represents a measurement of a property of system **S**, targeting the identification of the syndromes corresponding to the presence of a fault.

Let $\mathbf{T_S} = \{t_1, t_2, t_3 \ldots t_n\}$ denote the set of all tests used for system S, also referred to as *Test Suite*. In general, *testing* a system can be described as the operation of performing a set of tests [5]. In a diagnostic environment, tests are designed specifically to identify a potential syndrome of a failure, in order to make the defect observable within the system.

- **Definition 7**. A *syndrome* s characterizes the measurement of a test **t**.

Syndromes can provide detailed information about the output values of a circuit [8] or system parameters (e.g., the value of a sensor measurement [9]). In some cases, the information can be compacted in some form in order to keep only its most significant part, e.g., the correspondence between the observed value of the parameter and its expected one. When maximally compacted, a test outcome is a binary variable that takes the value *pass* or *fail*.

- **Definition 8**. A *root cause* is the faulty component that leads to the failure of system **S**.

The isolation of a root cause is achieved on the basis of the syndromes observed for some or for all tests $\mathbf{T_S}$. These syndromes contain the information that can be used to reveal the cause of misbehavior in system **S**.

A common assumption adopted in diagnosis is the so-called single fault hypothesis. This assumption is necessary in order to reduce the complexity of considering an exponential number of fault combinations to describe all possible system failures. According to this assumption, any instance of a system can contain at most one root-cause component.

## *1.1.2  Testing in the Manufacturing Line*

Manufacturing test ensures that electronic products have no defects before they are shipped to customers. Each fabricated product is subject to manufacturing test. The process of manufacturing test is described as the explanation or interpretation of a set of failure symptom [10], where a sequence of diagnostic tests is selected efficiently to locate the root causes of failures. A diagnosis strategy has to tackle the following fundamental issues [11]:

- **Detection**, i.e., the capability of a combination of tests to identify the presence of a fault in a system.
- **Isolation**, i.e., the capability of a diagnostic strategy to achieve fine-grained localization in order to allow the repair of a single component.

Compared to detection, isolation is a more difficult task in manufacturing test and it is the main focus of this book. We use the term *Diagnosis* to refer to root-cause isolation.

Diagnosis strategies vary across the different stages of system assembly. In order to better understand the problems and challenges at the board level, let us first examine chip-level testing. At the chip level, scan test is commonly used. In order to shift test patterns and response in and out of the chip, respectively, all the flip-flops are stitched to one or multiple scan chains. Test patterns are generated by automatic test-pattern generation (ATPG) tools and scanned into the circuit under test (CUT). The response of the circuit is compared with the expected response. The circuit is deemed to be fault-free if the test response matches the expected response. This testing process can be performed on automatic test equipment (ATE).

A modern ATE is a complex equipment operating at GHz frequency and with high throughput. The test cost on a high-end ATE can be up to thousands of dollars per pin [6]. Therefore, test-data compression techniques have been widely explored in the literature, in order to reduce the volume of test patterns and testing time. To reduce dependence of an expensive ATE, built-in-self-test (BIST) techniques have also been used [6]. In BIST, pseudo-random patterns are generated on-chip using a pattern generation circuit, e.g., a linear-feedback shift register (LFSR). The on-chip pattern generation approach eliminates the need for expensive external testers, and makes high-speed test possible. However, the reliance on pseudorandom patterns to achieve adequate fault coverage leads to a large volume of test data, compared to the use of deterministic patterns generated by ATPG tools.

Chip-level test has been well studied in the past. The testing goal is straightforward and quantifiable, i.e., to detect a high percentage of defects that are introduced during manufacturing. The test environment (ATE) is well-understood and a given number of pins are available for probing. Sophisticated tools and methodologies are available for inserting effective design-for-testability (DFT) structures and for automatically generating test patterns to get the desired fault coverage. This book focuses on the testing and diagnostic problems at higher integration levels. Much less attention has been devoted in the literature to these problems.

Circuit boards consist of previously tested components. An important objective of board testing is to verify the printed wiring and the contacts between wires and components. Moreover, the at-speed interactions between the components must also be tested. A typical manufacturing test line of electronic systems is shown in Fig. 1.1. Testing starts on the left-hand side of the figure. On the right-hand side, completely tested products are placed in inventory or shipped to customers. The testing process is separated into multiple stages to provide better failure isolation and feedback to the manufacturing process.

AOI              AXI                ICT            Functional Test          Burn-in Test

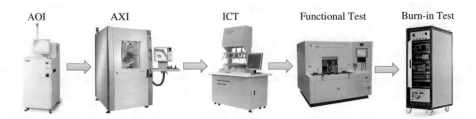

**Fig. 1.1** A typical manufacturing test line for electronic systems

Process test, such as automated optical inspection (AOI) and automated X-ray inspection (AXI), is first applied to immediately catch process flaws, e.g., solder shorts and unreliable solder joints. In-circuit test (ICT) is used to verify the performance of individual components using a bed-of-nails fixture. The bed-of-nails fixture is used to gain access to the board. ICT is useful to guarantee that a component receives the correct value, since many board failures are caused by open/short circuits or wrong components. Functional test, which targets the functional correctness of components and the whole system, is typically run after process tests. Sometimes, system test is performed as the final test, which is also a type of functional test. Each test technique has its advantages. For example, the solder reliability can be easily checked by AXI. The reversed or inoperative components can be detected by ICT. No single test can, however, cover all possible defects.

With the development of high-density assembly on PCBs, the number of access points for in-circuit test keeps decreasing. An interconnection test method, namely boundary-scan test, has been widely adopted by manufacturers. The boundary-scan architecture is defined in the IEEE 1149.1 and 1149.6 standards [12, 13] to ensure connectivity between components. Simple boards and complex multiboard systems can effectively be tested using the IEEE 1149.6 standard-compliant equipment from the product design phase to mainstream manufacturing. At present, there is an increased focus in the electronics industry on using the concept of remote test and diagnosis in order to provide a mechanism that allows continued support of a product [14]. The boundary-scan test architecture is illustrated in Fig. 1.2. A boundary-scan cell, which includes a multiplexer and latches, is added to each pin on the chip. Boundary-scan cells can capture data from pins or core logic signals, and force data to pins. The test data is serially shifted into the boundary-scan cells. Then, the captured data is serially shifted out and externally compared with the expected

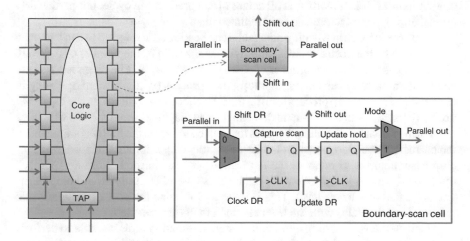

**Fig. 1.2** Illustration of the boundary-scan test architecture

responses. Boundary scan allows for full controllability and observability of board wires and therefore enables a simple interconnect test with a high fault coverage.

Despite the benefits offered by the above techniques, individual component testing and interconnection testing are no longer sufficient to guarantee the quality of complex circuit boards [15, 16]. It is often the case that all the chips on a board pass ATE tests, but the board still fails functional test, a scenario referred to as "No Trouble found" (NTF). The reason for NTF is that the board-level test environment is different from that at the chip level. At the chip manufacturing sites, chips are tested in a standalone mode, but additional issues, e.g., signal integrity, power-supply noise, and crosstalk involving multiple active chips, must be considered in board-level test [17]. The main objective of functional tests is to detect NTFs and verify the functional correctness of a board. Functional test plays an important role in capturing defects that cannot be easily detected by structural test.

Functional tests are indispensable today in system test and diagnosis [1]. Typically, functional test targets only a subset of all the designed functions, usually the critical functions identified by designers. Functional test sequences are often derived from design verification programs, and they are close to the practical scenarios that occur in the field. With the need for high throughput and efficient manufacturing, sophisticated functional test platforms based on open industry standards are required by electronics contract manufacturers and equipment manufacturers.

## 1.2  Introduction to Board-Level Diagnosis

Fault diagnosis isolates the root cause of a malfunction system by collecting and analyzing information on system status using measurements, tests, and other information sources. It is important at all stages of the product life cycle, but particularly crucial during manufacture and field maintenance.

The degree of accuracy with which faults can be located is referred to as diagnostic resolution [8]. Diagnostic success ratio refers to the ratio of the number of correctly diagnosed cases to the total number of cases under diagnosis. The diagnosis process can be hierarchically carried out as a top-down process (with a system operating in the field) or a bottom-up process (during the fabrication of a system). In the top-down process (system → boards → chips), the first-level diagnosis usually deals with large units such as boards. The faulty board is then tested to locate the faulty component on the board. Accurate location of faults inside a faulty chip is important information for chip manufacturers to improve the fabrication process. In the bottom-up approach (chips → boards → system), a higher level is assembled only from components already tested at a lower level. This is done to minimize the cost of diagnosis and repair, which escalates with the level at which faults are detected.

At the chip level, existing electronic design automation tools are able to perform accurate and high-resolution diagnosis. The failure mechanism, logic location, and even physical location of a fault can be determined, according to scan-test patterns [18]. The board-level diagnosis, in contrast, is much more challenging. There is no effective flow to locate the root cause of a failure during function tests. Once a failure is detected in the manufacturing line, the failed product is sent to the diagnosis department for repair. Typically, technicians run additional functional tests and measurements based on their personal experience. This process is time-consuming, and there is no guarantee of the success of repair. In addition, a board has to be scrapped after a few unsuccessful repair attempts. Current diagnostic software is not able to accurately and rapidly locate the root cause. Above all, the development of the diagnostic software is increasingly challenging, as electronic systems become more complex. Developing a diagnosis software relies heavily on the experience of debug technicians, which adds to total manufacturing costs.

## 1.2.1 Review of State-of-the-Art

A number of board-level fault diagnosis techniques have been presented in the literature [5, 9, 11, 19–25].

### 1.2.1.1 Rule-Based Methods

Rule-based diagnosis methods take the form "IF syndrome(s), THEN fault(s)" to locate a fault [19], as shown in Fig. 1.3. Hundreds or thousands of rules may be required to represent all the relevant knowledge for the system under diagnosis. Rule-based diagnosis involves the extraction of syndromes from the failure, and the firing of rules that match the syndromes. This process is repeated iteratively until the root cause of the failure is found. Rule-based expert systems have been developed for board repair and maintenance. The primary advantage of this flow is its simplicity and its ease of implementation for small systems, where rule-based methods can provide a powerful tool for quickly filtering out least likely hypotheses. These methods do not require a systematic understanding of the underlying system architecture or operational principles.

However, rule-based diagnosis systems suffer from a number of disadvantages that limit their use for complex systems. They do not incorporate adaptive learning

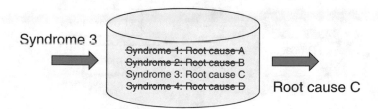

**Fig. 1.3** Illustration of rule-based diagnosis methods

from experience and they are unable to deal with unseen scenarios. This is correlated to the inability to update system knowledge. Furthermore, rule-based methods are inefficient in dealing with inaccurate information. In hierarchical systems, the lack of a reference with respect to the system structure makes it very difficult to reuse diagnosis system developed for similar systems.

Nevertheless, these approaches are important for historical reasons, since they represent the first attempts to solve diagnostic problems; furthermore, rules are the most immediate instrument to describe failure symptoms in the form of "IF syndrome(s), THEN fault(s)" relationship to produce a systematic description of engineering expertise.

For instance, in [20], the knowledge of the proposed rule-based system is a set of If-Then rules, connecting different observations of tests to diagnostic conclusions (fault candidates). The inference is performed by a sequential rule interpreter, which activates rules according to observations, until a unique diagnostic conclusion is reached.

In [26], authors reformulated fault diagnosis on the basis of a rule-based inference, which stores the test information as a matrix. Diagnosis conclusion implies a set of failures, and a failing test requires at least one diagnosis conclusion to be true.

### 1.2.1.2   Model-Based Methods

Model-based methods rely on an approximate representation of the system under diagnosis, as shown in Fig. 1.4. The system model is typically constructed in a hierarchical manner; higher level models are based on lower level models. The advantages of these methods are their robustness, straightforwardness, and ease of use. They have the potential to solve new problems and their knowledge may be organized in an expandable, upgradeable, and modular fashion.

However, a system model for complex topology and deep hierarchy is hard to obtain using model-based methods. In addition, it is difficult to develop efficient tests for these complex systems [22]. Due to design complexity, knowledge about

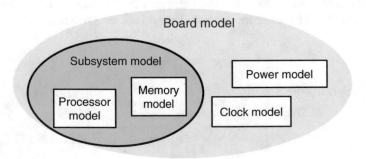

**Fig. 1.4**  Illustration of model-based diagnosis methods

the system is typically "spread" over different engineers. Integrating and testing these diagnosis systems is a time-consuming, tedious, and error-prone process.

Nevertheless, model-based diagnosis systems became popular during the last decade and they are still being utilized. For instance, [27] presents a comprehensive approach for model-based diagnosis, assuming that the system description is augmented with system structure.

In [23], Agilent Technologies presented a strategy to automate the debug process and allow for information sharing early in the design cycle between the hardware and software designers. Engineering information of a system model is documented during design time and it can be refined during product manufacturing time.

### 1.2.1.3  Reasoning-Based Methods

In reasoning-based diagnosis system, the fault isolation process is driven by an inference engine based on failure-symptoms correlation rules, expressed in the form of labels, as shown in Fig. 1.5. In particular, reasoning-based systems are a special class of expert systems that make their decisions based on experience and past situations. They attempt to acquire relevant knowledge from past cases and previously used solutions to propose solutions for new problems. This goal is achieved through the learning of correlation patterns. When a problem is successfully solved, the solution may be used in dealing with subsequent problems.

For example, in [28], authors proposed two diagnosis techniques based on fault dictionaries generated using fault signatures in the presence of failures. Here, fault-dictionary is a kind of reasoning engine. Dictionaries are used to train classifiers that can infer root causes from syndromes. One diagnosis system is designed as a fuzzy system to extract a set of if-then rules, while another is a radial-basis function system. In [29], Pous et al. proposed an approach based on distance metrics to overcome the limitation of the fault-dictionary approach. Since the faults used for building the dictionary may be different from the set of faults in the actual system under diagnosis, previously unrecognized faults may lead to incorrect prediction in actual diagnosis. Therefore, postprocessing is used to reduce ambiguity in the dictionary-based diagnosis system; however, the resulting computation may come at a high cost.

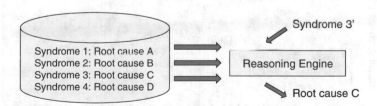

**Fig. 1.5** Illustration of reasoning-based diagnosis methods

Machine-learning-enabled reasoning diagnosis systems become prevailing in both industry and academy in recent decade. However, basic reasoning-based systems require an application-specific model and a large dataset of previous cases to avoid incorrect learning. Furthermore, a broad set of different failure manifestations is necessary to avoid overfitting [30]. From a computational point of view, time inefficiency may make them unusable in real-time diagnosis scenarios [26]. Details about machine-learning-enabled reasoning diagnosis systems are discussed in Sect. 1.2.3.

## 1.2.2  Automation in Diagnosis System

Reasoning-based methods are becoming popular today since detailed system models are not needed to construct the diagnosis system. Benefits of reasoning-based diagnosis system include:

- Reduced dependence on expert debug technicians.
- Faster ramp-up during the New Product Introduction (NPI) phase, since reasoning is available much sooner than the time needed by debug technicians to come up to speed (see Fig. 1.6).
- Transfer of diagnostic knowledge from engineering to manufacturing.

A reasoning-based diagnosis system is embedded in the manufacturing test line, as shown in Fig. 1.7. In today's manufacturing line, functional/system test is typically performed after products pass process test (AOI/AXI) and structural test (ICT and boundary scan test), in order to help isolation of the faults that escape structural tests. Functional test is designed to be close to the practical scenarios that can occur in the field. The conditions for each test are individually determined and set in order to mimic the particular practical scenario and not to influence the follow-up tests [32].

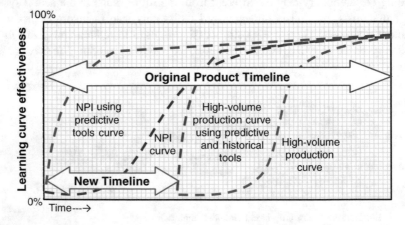

**Fig. 1.6**  Illustration of new product ramp-up with and without reasoning-based tools [31]

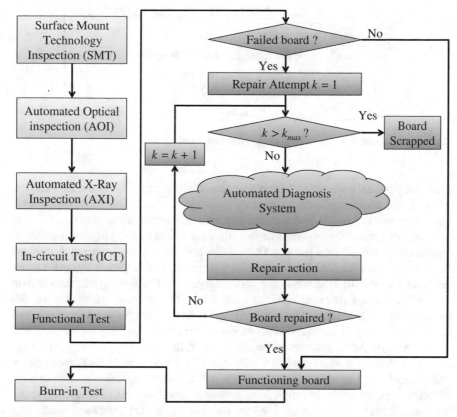

**Fig. 1.7** Flowchart for automated diagnosis

Once a failure is detected in the manufacturing line or in the field, the failed product is sent to the diagnosis department for repair. Typically, technicians run additional functional tests and measurements based on their personal experience. This process may take up to several days, and even several weeks if the board remains faulty after the component is replaced at the repair center [33]. In automated diagnosis system, test programs are loaded for diagnosing the faulty board. Based on the syndromes collected from the tests, the automated diagnosis system offers several suspicious root-cause candidates in a list, with the root cause of the most confidence on the top of the list. Repair action is executed on the suggested root-cause component. A repaired board is exercised again using test programs again to check functionality. If it continues to fail the test program, the faulty board will be taken back to the diagnosis system for another iteration of diagnosis. Note that a board has to be scrapped after a few unsuccessful repair attempts in a high-volume production environment. Compared to traditional diagnosis procedure executed by technician labors, which is time-consuming and no guarantee on the success of repair, the automated diagnosis system is more accurate and can be adaptive to the new knowledge. In addition,

```
SD587V_MASTERI:Interrupt Occur, SD587v_Masteri_Injt:0X00000028
SD587_srx_int:0x00000002
Interrupt caused by int_set[1]
SD587V_RBI_INT:0x00000020
A packet was received with a framing error in RBI
......
LA0 Engine-error status................0x1000 0000 00
LA1 Engine-error counter..............0x000f ffff
......
Error Code: Ox26
```

**Fig. 1.8** A segment of the log file for a failed functional test

current diagnostic software is not able to accurately and rapidly locate root causes. After all, the reasoning-based automatic diagnostic system can facilitate board-level fault diagnosis and repair, and significantly reduce manufacturing cost.

The extraction of fault syndromes is critical for model training in a diagnosis system that targets functional failures. Design-for-Testability (DFT) support is required for this purpose. Without appropriate DFT support, the diagnosis system cannot generate meaningful syndromes for board diagnosis. The fault syndromes should provide a complete description of the failure, and the extracted syndromes for different actions should have sufficient diversity such that ambiguity can be eliminated in the eventual repair recommendations. Fault syndromes vary across products and tests. Taking network systems as an example, traffic tests are performed through route processors, interface modules, and port ASICs to ensure correct functionality and connectivity. Critical fault syndromes in a failed traffic test often consist of error counters, drop counters, interrupt bits, mismatch interfaces, ECC/CRC errors, etc. To extract these fault syndromes, log files are parsed using keywords and syndromes are extracted using different descriptive-language scripts. For example, a segment of the log file for a failed traffic test is shown in Fig. 1.8. The fault syndromes extracted are int_set[1] (interrupt bit), LA0 Engine 0x0000ffff (error counter) and 0x26 (error code). Each of these elements is considered to be one syndrome. According to the syntax of different log files, changes may be needed in parsing scripts. The repair action is often directly recorded in the database, e.g., "replacement is component U37".

The above diagnosis flow requires the availability of a sufficient amount of fail data that can be used for training. If such data is not available in sufficient quantity, fault-insertion test (FIT) can be used to extract fail data in an artificial manner [16, 34]. FIT is a promising method for board/system reliability test and diagnosis coverage measurement. Due to the complexity of today's boards, random defects can occur, and unlike systematic defects that occur in a predictable manner and multiple times, random defects do not contribute in a meaningful way to system training. FIT verifies error detection by intentionally inserting wrong values at the pin/logic level. FITs can help mimic design corners and marginality, enrich the training set with rare events, and facilitate the timely release of a quality diagnosis system, and also provide feedback on the fault coverage of a complicated system.

### 1.2.3  New Directions Enabled by Machine Learning

Over the past decade, big data techniques (including machine learning, cloud computing, artificial intelligence, etc.) have been advocated in numerous fields, e.g., e-commerce, medicine, and the financial industry. Due to the nature of large volume production and multi-dimensional data collected in the electronics manufacturing line, manufacturing and product quality assurance is also an area where big data techniques can be applied. Fault diagnosis is one of the main machine-learning-enabled topics, that can contribute to the reduction of manufacturing cost.

In general, two main issues must be addressed when we attempt fault diagnosis for electronic systems, namely diagnosis accuracy and diagnosis speed.

#### 1.2.3.1  Research on Increasing Diagnosis Accuracy

A significant amount of academic and industrial research has focused on increasing diagnosis accuracy, since increasing diagnosis accuracy directly improves product yield. Butcher et al. [35] presented a diagnosis system based on Bayesian networks. This diagnosis system can be either derived from domain knowledge or learned from actual tests, which are extracted from debugging and maintenance data. A statistically significant (large) dataset is required to reflect an appropriate relationship between faults and test outcomes. However, since different components (with non-uniform failure rates) are used in different production lines, and equipment used for testing can vary, training a diagnosis system using these aggregated data may lead to incorrect learning. Therefore, correction techniques based on conditional probabilities are used to alleviate data corruption [35]. Similar Bayesian-network-based diagnosis systems can also be seen in [36, 37]. Since the relationship between root causes and syndromes is easy to be interpreted in Bayesian-network-based diagnosis systems, such diagnosis system is useful in many industrial applications [36, 37].

Artificial neural networks (ANNs) have also been used for fault diagnosis in digital systems. For instance, in [38], the authors built a diagnosing system using multilayer ANNs for combinational circuits. Training of this diagnosis system is based on a database of identified potential faults, which are generated by inserting one fault in the circuit followed by simulation. In [33], Zhang et al. proposed a single-layer ANN for board-level diagnosis to achieve a significant improvement over previous ANN-based diagnosis methods. However, the training time becomes prohibitively long as the system complexity increases.

Support-vector machines (SVM)-based diagnosis systems are also attracting attention. Zhang [33] and Ye [39, 40] proposed SVM-based diagnosis systems, which feature higher diagnosis accuracy over the existing Bayesian network and ANN-based diagnosis systems, because SVM-based diagnosis systems can efficiently avoid over-fitting and leverage the benefits of using multi-kernel techniques. In addition to fault diagnosis, Saunders et al. extended the use of SVMs to stock control and production planning for electronics manufacturing [41].

### 1.2.3.2   Research on Increasing Diagnosis Speed

Efforts have also been devoted to the analysis and ranking of tests to increase diagnosis speed [42, 43]. Sun [42] developed a diagnosis agent to facilitate evaluation of the quality of diagnosis tests by leveraging information about board structure and test coverage. This work aims to screen tests with weak diagnosis ability and increase the diagnosis ability at test-design stage. As a follow-up, Sun [43] also proposed a syndrome-merging framework with the help of domain knowledge in order to reduce the number of features in the diagnosis system, therefore increasing diagnosis speed. In this framework, syndrome evaluation is executed starting from early-design stage till volume production, such that the test sequence is dynamically evaluated and optimized from time to time.

Another diagnosis engine based on decision tree (DT) technique has also been widely studied and adopted [44, 45]. Ye [44] proposed DT-based diagnosis systems for board-level fault diagnosis. Faulty components are classified according to the discriminative ability of the syndromes in DT training. The diagnosis procedure is constructed as a binary tree, with the most discriminative syndrome as the root and final repair suggestions are available as the leaf nodes of the tree. The syndrome to be used in the next step is determined based on the observation of syndromes thus far in the diagnosis procedure. The number of syndromes required for diagnosis can be significantly reduced compared to the total number of syndromes used for system training. Moreover, in [45], online learning is facilitated in the proposed diagnosis system using an incremental version of DTs, so as to bridge the knowledge obtained at test-design stage with the knowledge gained during volume production.

### 1.2.3.3   Other Relevant Research

Work has also been carried out on model-update (i.e., refinement) solutions for reasoning-based diagnosis engines. An evaluation framework based on minimum-redundancy-maximum-relevance (mRMR) has been used to identify a set of syndromes with high discriminative ability [46–48]. However, it provides no diagnosis ability and requires additional algorithms for diagnosis. Bolchini et al. proposed Components-Tests Matrix (CTM) to evaluate the relationship between root causes and syndromes, such that the number of tests can be reduced for diagnosis [49].

Moreover, depending on the complexity of the product, it usually takes several months to accumulate an adequate database for training a reasoning-based diagnosis system. During the initial product ramp-up phase, reasoning-based diagnosis is not feasible for yield learning, since the required database is not available due to lack of volume. In [50], researchers proposed a knowledge-sharing mechanism to increase diagnosis accuracy in early-stage production.

More machine-learning-enabled diagnosis systems can be found in the literature [19, 21, 38, 40, 51–53].

## 1.2.4 Challenges and Opportunities

Printed circuit boards (PCBs) and electronic systems are found in a wide range of applications, e.g., vehicles, medical equipment, data centers, and network communications. With an increase in system complexity, the need for automated effective diagnostic tools has become more acute. This is driven by economic scaling of test, limited board test access, reduced time-to-market, shorter life cycles of products, and environmental requirements. We describe some of these issues in more detail below.

### 1.2.4.1 Cost of Manufacturing Test

With advances in process technology, the fabrication cost per transistor continues to decrease, but the testing cost per transistor remains the same or even increases slightly. According to the International Technology Roadmap for Semiconductors (ITRS) published in 2013 [54], the costs associated with testing and diagnosis is one of the highest contributors to manufacturing cost.

The cost of different levels of testing is often estimated using the rule of ten, as shown in Table 1.1 [55]. The table shows that if it costs $1 to test an ASIC, the cost of locating the same defective ASIC when it is mounted on a board is about $10; when the defective board is plugged into a system, the cost of identifying the faulty board and repairing the system is $100. If the defective system is delivered to end users, the cost of recall and repair will go up to $1,000. In 2007, the hardware failure on the Xbox 360, *the Red Ring of Death*, cost Microsoft more than $1 Billion [56]. In 2008, Amazon lost $31,000/min within 5 h of service outage [57]. Therefore, it is necessary to diagnose failures at an early stage of assembly and provide practical diagnostic tools to reduce diagnosis and repair cost.

### 1.2.4.2 Reduction of Yield Loss

The first-pass rate on current production lines is a matter of concern. For some complex telecommunication boards, only 60–70 % boards pass functional tests the first time [33]. The number of failed boards waiting for diagnosis accumulates rapidly, especially in high-volume production. Circuit boards have to be scrapped, if failures are still detected after a few times of repair. This is a significant impediment to profit, since a complex telecommunication board costs up to thousands of dollars [58]. Therefore, efficient and accurate diagnosis methods are urgently needed.

**Table 1.1** Rule of ten in test economics [55]

| Testing | ASIC testing | Board testing | System testing | Out in field |
|---|---|---|---|---|
| Cost of test | 1 | 10 | 100 | 1000 |

### 1.2.4.3 Accelerated Diagnosis

In an increasingly competitive marketplace, time-to-market is critical for the success of a company. This is another driver for the use of automated test and diagnostic tools. Isolating the root cause of board-level functional failures usually requires an expert, with engineering skills in both hardware and software. First, depending on the complexity of the product, it can take several months (even years) to develop this level of expertise. During initial product ramp-up, this expertise is especially needed but often unavailable, hence diagnosis preparation time is often very long [31]. Next, isolating the root cause itself takes long time, since debug technicians need to analyze the diagnosis log manually to identify the suspicious component. As the production volume increases and corresponding failing product volume increases, the diagnosis time inevitably increase dramatically. Therefore, an automated diagnosis engine is much needed for reducing the diagnosis time.

As discussed earlier in this chapter, a common scenario in industry is that of NTF. To address this problem, a diagnostic method is desired that can not only locate the faulty component on a malfunctioning board, but it can also narrow down the defective area inside the faulty component. The detailed failure information, e.g., failed test, failed submodules, failing cycles, is very important for component suppliers to troubleshoot and debug their manufacturing problems.

### 1.2.4.4 Accommodate New Assembly Technology

Environmental concerns, legislation, and market requirements are leading to increasing use of lead-free assembly technology in the electronics industry [59]. In 2006, the European Union Waste Electrical and Electronic Equipment Directive and Restriction of Hazardous Substances Directive (RoHS) came into effect to prohibit the intentional addition of lead to consumer electronics produced in the European Union. Quality and inspection of lead-free processes are emerging challenges in manufacturing test. The new assembly methods result in less access to boards for testing, debug and diagnosis. Changing package geometries are reducing the fault coverage that can be obtained using human visual inspection techniques. Some process imperfection and incorrect passive components may escape AXI, AOI, and ICT test. Detection and diagnosis of such faults have become new requirements in the functional test stage.

### 1.2.4.5 Expand Industry Adoption and Use

Another motivation for this line of research is that current application of automated board-level diagnostic tools is limited. Fault diagnosis has been an active area of recent research and development, but it still faces challenges related to accuracy, resolution, and large data requirements. Over the past decades, automating fault diagnosis using artificial intelligence (AI) techniques has been a major research topic. There has been much progress, but industry acceptance, particularly in cost-sensitive segments, has not been high. The techniques described in this book are expected to lead to generic and automated solutions for board-level fault diagnosis, which will provide quantifiable improvements over brute-force methods, trial-and-error manual repair, and previous proposed techniques based on expert systems. It is expected that the proposed automated solutions can be implemented and deployed in industrial manufacturing lines and debug departments.

## 1.3 Outline of Book

The remainder of this book is organized as follows.

Chapter 2 presents a fine-grained and adaptive diagnosis system for high-volume products. The proposed diagnosis system is based on multikernel support-vector machines (MK-SVMs) and incremental learning. The MK-SVM method leverages a linear combination of single kernels to achieve accurate faulty component classification based on the errors observed. The MK-SVMs thus generated can also be updated based on incremental learning, which allows the diagnosis system to quickly adapt to new error observations and provide even more accurate fault diagnosis.

Chapter 3 first reviews the use of artificial neural networks (ANNs). A data fusion technique is presented, namely majority-weight voting (MWV), to leverage multiple classifiers in our diagnosis system. The proposed MWV takes advantage of both ANNs and SVMs to provide an optimal repair-suggestion set.

Chapter 4 presents a decision tree-based diagnosis system to accelerate the diagnosis process. Considering that a large number of syndromes must be used in learning techniques to ensure diagnosis accuracy and effective board repair, the diagnosis cost can be prohibitively high due to the increase in diagnosis time and the complexity of syndrome collection/analysis. In the proposed adaptive decision tree-based diagnosis method, faulty components are classified according to the discriminative ability of the syndromes in DT training. The diagnosis procedure is constructed as a binary tree, with the most discriminative syndrome as root and final repair suggestions are available as the leaf nodes of the tree. The syndrome to be collected in the next step is determined based on the observations of syndromes collected thus far in the diagnosis procedure. The number of syndromes required for diagnosis can also be significantly reduced compared to the number of syndromes used for system training. Furthermore, an incremental version of DTs is used to facilitate online learning, so as to bridge the knowledge obtained at test-design stage with the knowledge gained during volume production.

Chapter 5 presents an informative evaluation and enhancement framework for a diagnosis system. Learning in a diagnosis system requires a rich set of test items (syndromes) and a sizable database of faulty boards. An insufficient number of failed boards, ambiguous root-cause identification, and redundant or irrelevant syndromes can render the diagnosis system ineffective. The proposed evaluation and enhancement framework is based on information-theoretic concepts for guiding diagnosis systems using syndrome and root-cause analysis. Syndrome analysis based on subset selection provides a representative set of syndromes with minimum redundancy and maximum relevance. Root-cause analysis measures the discriminative ability of differentiating a given root cause from others. The metrics obtained from the proposed framework can also provide guidelines for test redesign to enhance diagnosis.

Chapter 6 describes the design of preprocessing step in a diagnosis system to handle missing syndromes. Traditional diagnosis systems fail to provide appropriate repair suggestions when the diagnostic logs are fragmented and some error outcomes, or syndromes, are not available during diagnosis. Missing syndromes can be handled by using imputation methods. Several imputation methods are discussed and compared in terms of their efficiency in handling missing syndromes.

Chapter 7 presents a knowledge-discovery and a knowledge-transfer method for enhancing reasoning-based board-level functional fault diagnosis. Depending on the complexity of the product, it usually takes several months to accumulate an adequate database for training a reasoning-based diagnosis system. During the initial product ramp-up phase, reasoning-based diagnosis is not feasible for yield learning, since the required database is not available due to lack of volume. The proposed knowledge-discovery method and knowledge-transfer method can be used to facilitate board-level functional fault diagnosis. First, an analysis technique based on machine learning is used to discover knowledge from syndromes, which can be used for training a diagnosis engine. Second, knowledge from diagnosis engines used for earlier-generation products can be automatically transferred through root-cause mapping and syndrome mapping based on keywords and board-structure similarities.

Finally, Chap. 8 summarizes the contributions of the book.

# References

1. Vo T, Wang Z, Eaton T, Ghosh P, Li H, Lee Y, Wang W, Jun H, Fang R, Singletary D, Gu X (2006) Design for board and system level structural test and diagnosis. In: Proceedings IEEE international test conference (ITC), pp 1–10
2. Tourangeau S, Eklow B (2006) Test economics - what can a board/system test engineer do to influence supply operation metrics. In: Proceedings IEEE international test conference (ITC), pp 1–10
3. Chakraborty T, Chiang C-H, Van Treuren B (2007) A practical approach to comprehensive system test and debug using boundary-scan based test architecture. In Proceedings IEEE international test conference (ITC), pp 1–10
4. Backstrom D, Carlsson G, Larsson E (2005) Remote boundary-scan system test control for the ATCA standard. In: Proceedings IEEE international test conference (ITC), pp 788–797
5. Simpson WR, Sheppard JW (1994) System test and diagnosis. Springer, New York
6. Bushnell M, Agrawal V (2000) Essentials of electronic testing for digital, memory, and mixed-signal VLSI circuits. Springer, New York
7. Isermann R (2006) Fault-diagnosis systems: an introduction from fault detection to fault tolerance. Springer, New York
8. Abramovici M, Breuer MA, Friedman AD (1990) Digital systems testing and testable design, vol 2. Computer Science Press, New York
9. Isermann R (2005) Model-based fault-detection and diagnosis-status and applications. Annu Rev Control 29(1):71–85

10. Huang Y, McMurran R, Dhadyalla G, Jones RP (2008) Probability based vehicle fault diagnosis: Bayesian network method. J Intell Manuf 19(3):301–311
11. Simpson WR, Sheppard JW (1991) System complexity and integrated diagnostics. IEEE Des Test Comput. 8(3):16–30
12. IEEE Standard 1149.1 (2013) Test Access Port and Boundary-Scan Architecture. http://ieeexplore.ieee.org/stamp/stamp.jsp?tp=&arnumber=938734
13. IEEE Standard 1149.6 (2013) Boundary-Scan Tesing of Advanced Digital Networks. http://ieeexplore.ieee.org/stamp/stamp.jsp?tp=&arnumber=1196298
14. Reis I, Collins P, van Houcke M (2006) Online boundary-scan testing in service of extended products. In: Proceedings IEEE international test conference (ITC)
15. Parker K (2003) Defect coverage of boundary-scan tests: what does it mean when a boundary-scan test passes? In: Proceedings IEEE international test conference (ITC), pp 181–189
16. Eklow B, Hossein A, Khuong C, Pullela S, Vo T, Chau H (2004) Simulation based system level fault insertion using co-verification tools. In: Proceedings IEEE international test conference (ITC), pp 704–710
17. Conroy Z, Richmond G, Gu X, Eklow B (2005) A practical perspective on reducing ASIC NTFs. In: Proceedings IEEE international test conference (ITC)
18. Mentor Graphics (2013) Tessent Diagnosis. http://www.mentor.com/products/silicon-yield/products/diagnosis
19. Fenton W, McGinnity T, Maguire L (2001) Fault diagnosis of electronic systems using intelligent techniques: a review. IEEE Trans Syst Man Cybern Part C Appl Rev 31:269–281
20. Sydenham PH, Thorn R (2005) Handbook of measuring system design, vol 2. Wiley, Chichester
21. Amati L (2012) Test and diagnosis strategies for digital devices: methodologies and tools. PhD dissertation, Politecnico di Milano, Italy
22. Boumen R, Ruan S, de Jong I, Van De Mortel-Fronczak J, Rooda J, Pattipati K (2009) Hierarchical test sequencing for complex systems. IEEE Trans Syst Man Cybern Part A Syst Hum 39(3):640–649
23. Manley D, Eklow B (2002) A model based automated debug process. In: IEEE board test workshop, pp 1–7
24. Fang H (2011) Design-for-testability and diagnosis methods to target unmodeled defects in integrated circuits and multi-chip boards. PhD dissertation, Duke University, USA
25. Zhang Z (2011) Optimization of fault-insertion test and diagnosis of functional failures. PhD dissertation, Duke University, USA
26. Sheppard JW, Butcher S (2006) On the linear separability of diagnostic models. In: IEEE Autotestcon, pp 626–635
27. Darwiche A (1998) Model-based diagnosis using structured system descriptions. J Artif Intell Res 8:165–222
28. Catelani M, Fort A (2002) Soft fault detection and isolation in analog circuits: some results and a comparison between a fuzzy approach and radial basis function networks. IEEE Trans Instrum Meas 51(2):196–202
29. Pous C, Colomer J, Melendez J, de la Rosa JL (2003) Case base management for analog circuits diagnosis improvement. Case-based reasoning research and development. Springer, Berlin, pp 437–451
30. Hastie T, Tibshirani R, Friedman JJH (2001) The elements of statistical learning. Springer, New York
31. O'Farrill C, Moakil-Chbany M, Eklow B (2005) Optimized reasoning-based diagnosis for non-random, board-level, production defects. In: Proceedings IEEE international test conference (ITC), pp 173–179
32. Parvathala P, Maneparambil K, Lindsay W (2002) FRITS — a microprocessor functional BIST method. In: Proceedings IEEE international test conference (ITC), pp 590–598
33. Zhang Z, Chakrabarty K, Wang Z, Wang Z, Gu X (2011) Smart diagnosis: efficient board-level diagnosis and repair using artificial neural networks. In: Proceedings IEEE international test conference (ITC), pp 1–10

34. Zhang Z, Wang Z, Gu X, Chakrabarty K (2012) Physical-defect modeling and optimization for fault-insertion test. IEEE Trans VLSI Syst (TVLSI) 20(4):723–736
35. Butcher S, Sheppard JW (2009) Distributional smoothing in Bayesian fault diagnosis. IEEE Trans Instrum Meas 58(2):342–349
36. Najafi M, Auslander DM, Bartlett PL, Haves P (2008) Application of machine learning in fault diagnostics of mechanical systems. In: Proceedings of the international conference on modeling, simulation and control (ICMSC), pp 22–24
37. Zhang Z, Wang Z, Gu X, Chakrabarty K (2010) Board-level fault diagnosis using Bayesian inference. In: Proceedings IEEE VLSI test symposium (VTS), pp 1–6
38. Al-Jumah AA, Arslan T (1998) Artificial neural network based multiple fault diagnosis in digital circuits. In: Proceedings international symposium on circuits and systems (ISCAS), vol 2, pp 304–307
39. Ye F, Chakrabarty K, Zhang Z, Gu X (2014) Board-level functional fault diagnosis using multi-kernel support vector machines and incremental learning. IEEE Trans Comput-Aided Des Integr Circuits Syst (TCAD) 33(2):279–290
40. Ye F, Zhang Z, Chakrabarty K, Gu X (2013) Board-level functional fault diagnosis using artificial neural networks, support-vector machines, and weighted-majority voting. IEEE Trans Comput-Aided Des Integr Circuits Syst (TCAD) 32(5):723–736
41. Saunders C, Gammerman A, Brown H, Donald G (2000) Application of support vector machines to fault diagnosis and automated repair. In: Proceedings of the international workshop on principles of diagnosis, pp 1–5
42. Sun Z, Jiang L, Xu Q, Zhang Z, Wang Z, Gu X (2013) AgentDiag: an agent-assisted diagnostic framework for board-level functional failures. In: Proceedings IEEE international test conference (ITC), pp 1–8
43. Sun Z, Jiang L, Xu Q, Wang Z, Zhang Z, Gu X (2015) On test syndrome merging for reasoning-based board-level functional fault diagnosis. In: Proceedings IEEE Asia South Pacific design automation conference (ASP-DAC), pp 737–742
44. Ye F, Zhang Z, Chakrabarty K, Gu X (2012) Adaptive board-level functional fault diagnosis using decision trees. In: Proceedings IEEE Asian test symposium (ATS), pp 202–207
45. Ye F, Chakrabarty K, Zhang Z, Gu X (2016) Adaptive board-level functional fault diagnosis using incremental decision trees. IEEE Trans Comput-Aided Des Integr Circuits Syst (TCAD) 35(2):323–336
46. Ye F, Zhang Z, Chakrabarty K, Gu X (2014) Information-theoretic syndrome and root-cause analysis for guiding board-level fault diagnosis. In: Proceedings IEEE European test symposium (ETS), pp 1–6
47. Ye F, Chakrabarty K, Zhang Z, Gu X (2014) Information-theoretic framework for evaluating and guiding board-level functional-fault diagnosis. IEEE Des Test Comput 31(3):65–75
48. Ye F, Zhang Z, Chakrabarty K, Gu X (2015) Information-theoretic syndrome evaluation, statistical root-cause analysis, and correlation-based feature selection for guiding board-level fault diagnosis. IEEE Trans Comput-Aided Des Integr Circuits Syst (TCAD) 34(6):1014–1026
49. Bolchini C, Quintarelli E, Salice F, Garza P (2013) A data mining approach to incremental adaptive functional diagnosis. In: Proceedings IEEE international symposium on defect and fault tolerance in VLSI systems (DFT), pp 13–18
50. Ye F, Zhang Z, Chakrabarty K, Gu X (2014) Knowledge discovery and knowledge transfer in board-level functional fault diagnosis. In: Proceedings IEEE international test conference (ITC), pp 1–10
51. Wang H, Poku O, Yu X, Liu S, Komara I, Blanton RD (2012) Test-data volume optimization for diagnosis. In: Proceedings ACM/IEEE design automation conference (DAC), pp 567–572
52. Bolchini C, Cassano L (2014) Machine learning-based techniques for incremental functional diagnosis: a comparative analysis. In: Proceedings IEEE international symposium on defect and fault tolerance in VLSI systems (DFT), pp 246–251
53. Rajan V, Yang J, Chakrabarty S, Pattipati K (1998) Machine learning algorithms for fault diagnosis in analog circuits. IEEE international conference on systems, man, and cybernetics, vol 2, pp 1874–1879

54. International Technology Roadmap for Semiconductors (ITRS'13) (2013) http://www.itrs.net/
55. Lin Y-T (2005) Economic designs for manufacturing system test and field maintenance. In: Proceedings Southeastern symposium on system theory, pp 40–44
56. Xbox 360 technical problems "Ring of Death" (2007) http://en.wikipedia.org/wiki/Xbox_360_technical_problems
57. Outages hit Amazon's S3 storage service (2008) http://www.networkworld.com/news/2008/072108-amazon-outages.html
58. Cisco Line Cards (2013) http://www.cisco.com/en/US/products/hw/modules/ps2710/prod_module_series_home.html
59. iNEMI Technolog Roadmaps, from International Electronics Manufacturing Initiative Organization (2009) http://www.inemi.org/node/2151

# Chapter 2
# Diagnosis Using Support Vector Machines (SVM)

Diagnosis of functional failures at the board level is critical for improving product yield and reducing manufacturing cost. State-of-the-art board-level diagnostic software is unable to cope with high complexity and ever-increasing clock frequencies, and the identification of the root cause of failure on a board is a major problem today. Ambiguous or incorrect repair suggestions lead to long debug times and even wrong repair actions, which significantly increase the repair cost and adversely impacts yield.

In this chapter, we introduce a machine learning-based intelligent diagnosis system, which can automatically learn debug knowledge from empirical data and identify the most likely root cause of a new failed board. Using such a diagnosis system eliminates the difficulties involved in traditional knowledge acquisition. Fine-grained fault syndromes extracted from failure logs and the corresponding repair actions are used to train the system. Support vector machines (SVMs) have been used in board-level diagnosis to provide accurate root cause isolation. An SVM-based diagnosis system can be rapidly trained and is scalable to large datasets. However, the SVM method used in prior work [1] was simplistic, relying on an arbitrarily chosen kernel function, and it was not adaptive to the availability of new data or test cases. We propose a diagnosis system based on multi-kernel support vector machines (MK-SVMs) and incremental learning, which are used to tune the diagnosis system in an automatic manner. The MK-SVM method leverages a linear combination of single kernels to achieve accurate faulty component classification based on the errors observed. The MK-SVMs thus generated can also be updated based on incremental learning, which allows the diagnosis system to quickly adapt to new error observations and provide even more accurate fault diagnosis.

The remainder of this chapter is organized as follows. Section 2.1 reviews the background and prior work. Section 2.2 reviews basic concepts in support vector machines. Section 2.3 introduces multi-kernel-based SVMs, and describes how MK-SVMs can be extended for incremental learning, namely iMK-SVMs. Section 2.4 presents experimental results on diagnosis accuracy and training time for two industry

© Springer International Publishing Switzerland 2017
F. Ye et al., *Knowledge-Driven Board-Level Functional Fault Diagnosis*,
DOI 10.1007/978-3-319-40210-9_2

boards and for synthetic data. These results are compared to diagnosis using single-kernel SVMs [1] and ANNs [2]. In addition, experimental results are presented for the diagnosis accuracy achieved using incremental learning. The high diagnosis accuracy, rapid training, and short diagnosis time highlight the benefits of the iMK-SVM-based reasoning system. Section 2.5 concludes the chapter.

## 2.1  Background and Chapter Highlights

Field data and experience reports from repair technicians highlight many problems with diagnostic software currently in use, especially for functional tests that involve actual data in a real application. The diagnostic resolution offered by today's tools is limited to ASICs on the board. No repair guidance is provided for memory devices or passive components on the board. Diagnostic resolution is also poor in practice, multiple repair candidates are often listed and these candidates are not prioritized. Technicians are forced to run debug programs repeatedly and carry out physical probing in many places to identify the root cause of failures, a practice that significantly increases the debug and repair time. Based on past repair records, we have found the debug time for the functional test considered here to be as high as several weeks. The correctness of diagnosis, i.e., the probability of the actual failing component included in the list of suspects, is unacceptably low, and the root cause is seldom exclusively pinpointed.

In order to overcome the difficulties described above and provide accurate diagnostic results, we investigate intelligent diagnosis based on machine-learning algorithms. Machine learning, a branch of artificial intelligence, is focused on automatic learning from empirical data and making intelligent decisions. The debug knowledge can be automatically learned from history records (logs) using these techniques, e.g., artificial neural networks (ANNs) [2]. In ANN learning, we are given a set of training cases, which typically contain a set of error observations, referred to as syndromes. An ANN aims to automatically generate both the edge weights in the network and a transfer function that allow root-cause identification to be made on the basis of the syndromes. Due to its wide acceptance in the machine learning community and ease of interpretation, ANNs have been used for fault diagnosis [2, 3]. However, ANN-based methods suffer from the inherent theoretical limitations of ANNs that tend to limit their accuracy [2]. Moreover, ANNs require large datasets for training, and large volumes of relevant data are not always available.

Recently, success in board-level functional fault diagnosis has also been reported using SVMs, which constitute a more advanced class of machine-learning techniques [1]. Even though SVMs were shown to be more effective than ANNs for a complex board in high-volume production [1], the increase in success ratio (diagnosis accuracy) was marginal. Moreover, the SVM method used in [1] was simplistic, relying on an arbitrarily chosen kernel function, and it was not adaptive to the availability of new data or test cases.

In this chapter, we propose an adaptive, accurate, and efficient diagnostic system based on SVMs. An advantage of using machine learning is that it avoids the difficulties associated with knowledge acquisition and rule-base development required for expert systems [1, 2, 4]. Without the need to understand the complex functionality of boards, diagnostic systems based on machine learning are able to automatically derive and exploit knowledge from repair logs of previously documented cases. The proposed approach overcomes the limitations of single-kernel SVMs used in [1] by exploiting multi-kernel SVMs and incremental learning (iMK-SVM) to reduce complexity, achieve significantly higher diagnosis accuracy, and perform reasoning adaptively in realtime as new data becomes available. The kernel function in this approach is defined as a linear combination of different kernels. The proposed iMK-SVM-based diagnostic system is generic. Given a set of fault syndromes and the corresponding faulty components, the system can be rapidly trained and then used for fault diagnosis across different products. Results are presented for two industry boards, which are currently in production, and for which fail data has been gathered and used for training and evaluation.

## 2.2 Diagnosis Using Support Vector Machines

A SVM is a supervised machine learning algorithm proposed by Vapnik in 1995 [5]. It has a number of theoretical and computational merits, for example, the simple geometrical interpretation of the margin, uniqueness of the solution, statistical robustness of the loss function, modularity of the kernel function, and overfitting control through the choice of a single regularization parameter. A brief introduction to SVMs is presented below.

### 2.2.1 Support Vector Machines

The goal of SVMs is to define an optimal separating hyperplane (OSH) to separate two classes. The vectors from the same class fall on the same side of the optimal separating hyperplane, and the distance from the closest vectors to the optimal separating hyperplane is the maximum among all the separating hyperplanes. An important and unique feature of this approach is that the solution is only based on those vectors that are the closest to the OSH, calculated in the following way. Let $(x_i, y_i)$, $i = 1, 2, \ldots, n$ be a set of training examples, and $x_i \in R^d$, $y_i \in \{-1, +1\}$. Figure 2.1 illustrates a two-class SVM model. The vector $x_i$ is considered as input, and $d$ is the dimensionality of the input vectors. Each input vector belongs to one of the two classes. One is labeled by $y = +1$; the other is labeled by $y = -1$. If the set can be linearly separated, there must be a hyperplane satisfying Formula (2.1):

$$f(x) = sgn(\omega^T x + b), \tag{2.1}$$

**Fig. 2.1** Illustration of a
2-class support vector
machine model

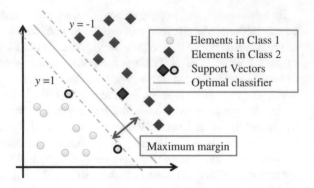

where *sgn* refers to the sign of $(\boldsymbol{\omega}^T \boldsymbol{x} + b)$, $\boldsymbol{\omega}$ is a $d$-dimensional vector, and $b$ is a scalar. Those vectors $\boldsymbol{x}_i$ for which $f(\boldsymbol{x}_i)$ is positive are placed in one class, while vectors $\boldsymbol{x}_i$ for which $f(\boldsymbol{x}_i)$ is negative are placed in another class. Based on [5], we define *margin* as twice the distance from the classifier to the closest data vector, namely the *support vector*. The margin is a measure of the ability to generate a classifier. The larger the margin is, the better is the generation of the classifier. SVMs maximize the margin between two classes.

Since the margin width equals $\frac{2}{\sqrt{\boldsymbol{\omega}^T \boldsymbol{\omega}}}$, the maximum-margin solution is found by solving the following minimization problem:

$$\text{Minimize} \quad W = \frac{1}{2} \|\boldsymbol{\omega}\|^p + C \sum_{i}^{S} \xi_i \tag{2.2}$$

Subject to

$$\|y_i(\boldsymbol{\omega} \cdot \boldsymbol{x}_i + b)\| \leq 1 - \xi_i, \ \forall i \tag{2.3}$$

$$\xi_i > 0, \ \forall i \tag{2.4}$$

where slack variable $\xi_i$ is introduced to measure the degree of misclassification of data $\boldsymbol{x}_i$ and $C$ is the error penalty. We can tune $C$ to adjust the trained SVM model to be either overfitting or underfitting. The parameter $p$ is used for regularization of the weights in the SVM model. Most SVM solvers use standard regularization, i.e., $p = 2$. Hence we assume $p = 2$ in our work.

In order to solve the constrained optimization problem described in (2.2), a set of Lagrange multipliers $\alpha_i$, where $\alpha_i \geq 0$, is used. Each multiplier $\alpha_i$ corresponds to a constraint in (2.3) on the support vectors. The optimization problem from (2.2) can now be expressed in its dual form

$$\text{Minimize} \quad W_1 = \frac{1}{2} \sum_{i,j=1}^{S} \alpha_i Q_{ij} \alpha_j - \sum_{i=1}^{S} \alpha_i + b \sum_{i=1}^{S} y_i \alpha_i \tag{2.5}$$

where $Q_{ij} = y_i y_j K(x_i, x_j)$, and $K$ is the kernel function described in the next section. Additional mathematical details are omitted here but they can be found in [5]. The weights and offsets are as follows:

$$\omega = \sum_{i=1}^{S} \alpha_i y_i x_i \tag{2.6}$$

$$b = \frac{1}{N_S} \sum_{i=1}^{S} (y_i - \omega \cdot x_i) \tag{2.7}$$

Originally, SVMs were designed for linear binary classification problems. In practice, classification problems are not limited by two classes. In board-level fault diagnosis, the number of root cause candidates (classes) is typically in the range of a few hundreds. In [5], a modified design of SVMs was proposed in order to incorporate multiclass learning. Besides this, an alternative approach for handling a large number of classes is to combine multiple binary SVMs. "One against one" provides pairwise comparisons between classes. "One against all" compares a given class with all the other classes. According to a comparison study in [5], the accuracies of these methods are almost the same. Therefore, we choose the "one against all" in our problem, which has the lowest computation complexity.

### 2.2.1.1 Demonstration of SVM-Based Diagnosis System

To illustrate the SVM optimization procedure, consider the same hypothetical demonstration board with six cases in Sect. 2.2. We build an SVM model to identify faults for new cases. Let $x_1$, $x_2$, and $x_3$ be three syndromes. If the syndrome manifests itself, we record it as 1, and 0 otherwise. The presentation of fault class is different from that for ANNs training in Sect. 2.2. The board has two candidate root causes A and B, and we encode them as $y = -1$ and $y = 1$, respectively. In a real scenario, fault syndromes vary across products and tests. Here, we merge the syndromes and the known root causes into one matrix $\mathscr{A}' = [\mathscr{B}'|\mathscr{C}']$, where the left ($\mathscr{B}'$) side refers to syndromes, while the right side ($\mathscr{C}'$) refers to the corresponding fault classes. This matrix represents the training information for the SVM.

$$\mathscr{A}' = [\mathscr{B}'|\mathscr{C}'] = \begin{bmatrix} 1 & 1 & 0 & \vdots & 1 \\ 1 & 1 & 1 & \vdots & 1 \\ 1 & 1 & 0 & \vdots & 1 \\ 0 & 1 & 1 & \vdots & -1 \\ 0 & 0 & 1 & \vdots & -1 \\ 0 & 0 & 1 & \vdots & -1 \end{bmatrix} \tag{2.8}$$

We obtain the Lagrange multipliers $\alpha_1 = 2.00e^{-7}$, $\alpha_2 = 1.99$, $\alpha_3 = 1.99$, and $\alpha_4 = 1.99e^{-7}$ by solving the optimization (cost) function from (2.5). We then get

$\omega_1 = 1.99$, $\omega_2 = 0$, $\omega_2 = 0$ and $b = -1.00$ by solving Eqs. (2.6) and (2.7). Therefore, the classifier for determining the root cause for the given set is generated as follows:

$$f(\boldsymbol{x}) = sgn(1.99 \cdot x_1 + 0 \cdot x_2 + 0 \cdot x_3 - 1.00) \tag{2.9}$$

Next, suppose a new failing board is received and it has the syndrome [1 1 0], which corresponds to the first row (case) of $\mathscr{A}'$ in Eq. (2.8). The function $y$ is evaluated using Eq. (2.2), and since $sgn(1.99 \cdot 1 + 0 \cdot 1 + 0 \cdot 0 - 1.00)$ is positive, $y = 1$. Thus the root cause for this failing board is determined to be A. Suppose a second new failing board with syndrome [0 1 0] in received. In this case, the decision function evaluates to $y = -1$, hence we determine B to be the root cause in this case. For boards with the root cause of class A (B), we can replace the corresponding component A (B).

### 2.2.2  SVM Diagnosis Flow

The SVM-based diagnosis flow consists of four steps that are described in Fig. 2.2. Generally speaking, a set of training data (fault syndromes and corresponding repair actions) is first prepared, which is derived from the repair history in the manufacturing database. Then SVMs determine the OSH based on the training data. After the OSH is determined, the SVMs-based diagnostic system is ready to diagnose new cases.

**Step 1**: The data preparation step also follows the description in Sect. 2.1. The extracted syndromes and replaced components are used as inputs and outputs for the training of SVMs.

**Step 2**: Proper kernel function and penalty parameter are chosen to determine the SVMs training. The choice of these two parameters affects the performance of SVMs.

**Fig. 2.2** The diagnosis flow using SVMs [1]

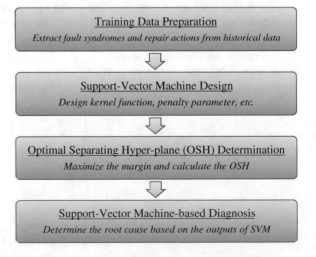

However, there are no generic rules to select the best kernel and other parameters of SVMs for a specific problem. In this work, we determine the best design of SVMs in a heuristic way. According to extensive experimental results, we find that the SVMs with a linear kernel and a relatively large penalty parameter provide the highest diagnostic accuracy in the board-level fault diagnosis.

**Step 3**: The determination of the OSH can be considered as the training of SVMs. The OSH is determined by solving the quadratic optimization problem described in Eqs. (2.7) and (2.6). The values of $w$ and $b$ can be calculated using MATLAB. An open-source SVM toolbox is provided in [6]. The $w$ and $b$ values are determined after training.

**Step 4**: Given a new input vector, we can calculate the output of the SVM using the decision function in Eq. (2.1). In the diagnosis step, we rank the output of all the SVMs, and select the component represented by the SVM with the largest output as the root cause.

## 2.3 Multi-kernel Support Vector Machines and Incremental Learning

Classical SVMs are efficient for linear classification, as discussed in Sect. 2.2. However, in many practical scenarios, including fault diagnosis, classical SVMs fail to find an optimal linear classifier for separating classes. Therefore, in such scenarios, SVMs must be extended to handle nonlinear classification problems. One solution is to transform the problem to a higher dimensional feature space through a nonlinear mapping, also known as *kernel*, and the classifier is constructed in the new feature space. The advantage of this transformation is that it is not necessary to explicitly implement the transformation and to determine the separating hyperplane in a higher dimensional feature space.

### 2.3.1 Multi-kernel Support Vector Machines

#### 2.3.1.1 Kernel

In kernel-based transformation methods, the data representation is implicitly chosen through a *kernel* $K(x_i, x_j)$, where $x_i$ and $x_j$ are both input vectors in the lower dimensional feature space. Figure 2.3 illustrates the transformation from a lower dimension feature space to a high-dimension feature space. The optimization problem from (2.5) can now be expressed as:

$$\text{Minimize} \quad W_2 = \frac{1}{2} \sum_{i,j=1}^{S} \alpha_i \alpha_j y_i y_j K(x_i, x_j) - \sum_{i=1}^{S} \alpha_i + b \sum_{i=1}^{S} y_i \alpha_i \qquad (2.10)$$

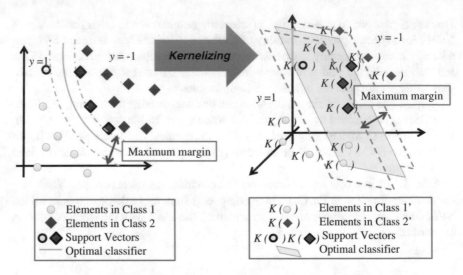

**Fig. 2.3** Illustration of kernelized support vector machine model

with similar constraint functions as those for nonkernel SVMs.

$$b = \frac{1}{N_S} \sum_{i=1}^{S} \left( y_i - \sum_{j=1}^{S} \alpha_j y_j \cdot K(\boldsymbol{x}_i, \boldsymbol{x}_j) \right) \tag{2.11}$$

The decision function (2.1) is now expressed as

$$f(\boldsymbol{x}) = sgn\left( \sum_{i=1}^{S} \alpha_i y_i K(\boldsymbol{x}, \boldsymbol{x}_i) + b \right) \tag{2.12}$$

The choice of kernel function is crucial for the success of an SVM-based model. There are several widely used kernel functions [5]:

- Homogeneous Polynomial Kernel: $K(\boldsymbol{x}_i, \boldsymbol{x}_j) = (\boldsymbol{x}_i \cdot \boldsymbol{x}_j)^d$, where $d \geq 1$. The linear kernel ($d = 1$) is regarded as a kernelized representation of linear SVMs.
- Polynomial kernel: $K(\boldsymbol{x}_i, \boldsymbol{x}_j) = (\boldsymbol{x}_i \cdot \boldsymbol{x}_j + 1)^d$, where $d \geq 1$.
- Gaussian kernel: $K(\boldsymbol{x}_i, \boldsymbol{x}_j) = e^{-\gamma \|\boldsymbol{x}_i - \boldsymbol{x}_j\|^2}$, where $\gamma = \frac{1}{2\sigma^2}$ and $\sigma$ can be interpreted as the standard deviation of a Gaussian distribution.
- Exponential kernel: $K(\boldsymbol{x}_i, \boldsymbol{x}_j) = e^{-\gamma \|\boldsymbol{x}_i - \boldsymbol{x}_j\|}$.

As an illustration, we apply kernelization methods to the example described in Sect. 2.2. Suppose that we choose a polynomial kernel with degree 2 and $K(\boldsymbol{x}_i, \boldsymbol{x}_j) = (\boldsymbol{x}_i \cdot \boldsymbol{x}_j + 1)^2$. Given a new failing board under test with syndrome [1 1 1], the kernelized presentation of the syndrome is [0.91 0.64 0.64] and $f(\boldsymbol{x})$ is positive. Therefore, we can classify it as being in class A, i.e., the root cause is A. Suppose another new failing board is received with syndrome [0 1 0]; $f(x)$ is now negative. Hence, this board is classified as being in class B, i.e., the root cause of failure for this board is B.

### 2.3.1.2 Multi-kernel Support Vector Machines

Recent applications in bioinformatics have shown that using multiple kernels instead of a single one can lead to better classification [7, 8]. The key idea here is to represent the kernel $K(x_i, x_j)$ as a linear combination of $M$ basis kernels:

$$K(x_i, x_j) = \sum_{k=1}^{M} \mu_k K_k(x_i, x_j),$$  (2.13)

where $\mu_k \geq 0$ and $\sum_{k=1}^{M} \mu_k = 1$. Each basis kernel $K_k$ can be one of the kernel types listed in Sect. 2.3. Thus the optimization problem can now be stated as

$$\text{Minimize} \quad W_3 = \frac{1}{2} \sum_{i,j=1}^{S} \alpha_i \alpha_j y_i y_j k \sum_{k=1}^{M} \mu_k K_k(x_i, x_j)$$

$$- \sum_{i=1}^{S} \alpha_i + b \sum_{i=1}^{S} y_i \alpha_i,$$  (2.14)

The optimization problem of Eq. (2.14) is solved using a reduced gradient method, the details of which are described in [9]. The training mechanism of a multi-kernel SVM-based diagnosis system is illustrated in Fig. 2.4.

In previous work [1], the SVM-based diagnosis system leverages single kernels in a heuristic manner. However, due to the correlation between syndromes, we cannot arbitrarily determine a single kernel for each diagnosis system. A diagnosis system requires an adaptive kernel in order to achieve higher prediction accuracy. Such adaptation can extend from a single kernel to multiple kernels. Without knowledge of the exact kernel to be used, the weights of different kernels can be appropriately configured to fit the training data and generate better prediction for the diagnosis system.

## 2.3.2 Incremental Learning

Incremental learning can not only solve the SVM training problem for large-scale data sets, but it can also facilitate online learning for SVMs as new data for failing boards and the corresponding repair outcomes become available. The use of multi-kernel SVMs increases computational complexity and diagnosis solutions take longer to converge. This problem can be tackled using incremental learning. By distributing

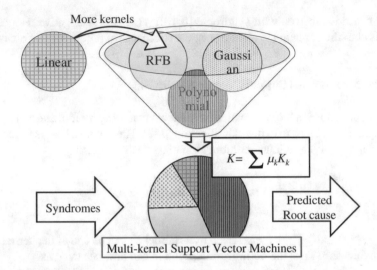

**Fig. 2.4** Illustration of training a multi-kernel SVM

the computational workload among different epochs during training, computational complexity can be reduced in terms of both computing time and memory allocation. Since SVM models can be determined on the basis of support vectors only, and the number of support vectors is typically very small compared to the number of training examples, SVMs can benefit from incremental learning through the compression of data from previous batches in the form of their support vectors. This approach to incremental learning with SVMs has been investigated in [10], where it has been shown that incrementally trained SVMs are as effective as their nonincrementally trained equivalents. Incremental SVMs can be described as the following optimization problem as an extension to (2.2):

$$\text{Minimize} \quad W^\diamond = \frac{1}{2}\|\boldsymbol{\omega}\|^2 + C\left(L\sum_{i\in S^*}\xi_i + \sum_{i\in S}\xi_i'\right) \tag{2.15}$$

with the same constraints (2.3) and (2.4). The parameter $S^*$ denotes the set of existing support vectors extracted from the previous SVM models, and $S$ is the new training set. As an optimization knob, the use of existing support vectors can be penalized by $L$ to model the fact that an erroneous decision made on the basis of previous support vectors is more costly than an error on a new example based on the current data. Incremental learning can be made more effective by combining it with multiple kernels, an approach that we refer to as iMLK-SVMs. A flowchart for this procedure is shown in Fig. 2.5.

As an illustration, consider an existing diagnosis system given by Eq. (2.1) and based on the input cases (boards) represented by the matrix $\mathscr{A}$ in Eq. (2.8). The support vectors in the existing diagnosis system can be extracted as shown in $\mathscr{A}'$.

**Fig. 2.5** Illustration of the proposed iMK-SVM approach for fault diagnosis

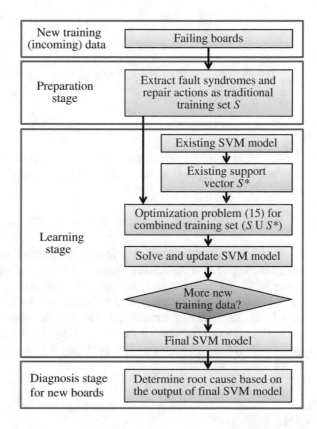

Suppose that we are given four new training boards (cases) indicated by the matrix $\mathcal{N}$ in Eq. (2.3.2) for incremental learning. The classifier (2.1) gives inconsistent root-cause identification—root cause A with the existing data but B with the new data. Therefore, we form a new data set $\mathcal{D} = [\mathscr{A}'; \mathcal{N}]$ as shown in Eq. (2.3.2).

$$\mathscr{A}' = \begin{bmatrix} 1 & 1 & 0 & \vdots & 1 \\ 1 & 1 & 1 & \vdots & 1 \\ 0 & 1 & 1 & \vdots & -1 \\ 0 & 0 & 1 & \vdots & -1 \end{bmatrix}$$

$$\mathcal{N} = \begin{bmatrix} 0 & 1 & 0 & \vdots & 1 \\ 0 & 1 & 0 & \vdots & 1 \\ 1 & 1 & 1 & \vdots & -1 \\ 1 & 0 & 1 & \vdots & -1 \end{bmatrix}$$

$$\mathcal{D} = \left[ \begin{array}{c} \mathcal{A}' \\ \hline \mathcal{N} \end{array} \right] = \left[ \begin{array}{ccc|c} 1 & 1 & 0 & 1 \\ 1 & 1 & 1 & 1 \\ 0 & 1 & 1 & -1 \\ 0 & 0 & 1 & -1 \\ \hline 0 & 1 & 0 & 1 \\ 0 & 1 & 0 & 1 \\ 1 & 1 & 1 & -1 \\ 1 & 0 & 1 & -1 \end{array} \right]$$

We obtain a unique solution to the new optimization problem from (2.15). We then get $\omega_1 = 0$, $\omega_1 = 0$, $\omega_1 = -2.00$ and $b = 1.00$. The classifier is now updated from (2.1) to be:

$$f(\boldsymbol{x}) = sgn(0 \cdot x_1 + 0 \cdot x_2 - 2.00 \cdot x_3 + 1.00) \tag{2.16}$$

Next, we suppose that a new failing board is received and has the syndrome [1 1 1], which corresponds to the third row (case) of $\mathcal{N}$. The function $f(\boldsymbol{x})$ is now evaluated using Eq. (2.16) and its value is negative, rather than the positive prediction based on the previous decision function of (2.1). Therefore, we can conclude that the failing board is in Class B, i.e., the root cause is B. Suppose another failing board is received with syndrome [1 1 0], which is the first test case from Sect. 2.2. In this case, the new decision function still evaluates $f(\boldsymbol{x})$ to be positive; hence, we determine A to be the root cause for this board.

## 2.4  Results

Experiments were performed on two industrial boards that are currently in high-volume production. Relevant information about the boards is provided in Table 2.1. For training, a total of 1613 repaired boards are collected from the contract manufacturer's database for Board 1. A total of 546 fault syndromes are extracted from failure logs. The number of faulty components identified in the database for repair action is 153. For Board 2, a total of 1351 repaired boards are analyzed as training data. A total of 375 fault syndromes are extracted from failure logs. The number of faulty components for repair action is 116.

**Table 2.1**  Information about the industrial boards used for classification and the log data available

|                                          | Board 1 | Board 2 |
| ---------------------------------------- | ------- | ------- |
| Number of syndromes                      | 546     | 375     |
| Number of repair candidates (components) | 153     | 116     |
| Number of boards                         | 1613    | 1351    |

The SVM algorithms are implemented using the MATLAB 2010b toolbox. Multi-kernel SVMs are implemented using SimpleMKL [9]. Incremental learning is implemented using McpIncSVMs [11] and the SVM-KM toolbox [6]. As a comparison, the ANN method described in [2] has also been implemented using the Neural Network toolbox in MATLAB 2010b [12]. Experiments were run on a 64-bit Windows system with 12 GB of RAM and quadcore Intel i7 processors running at 2.67 GHz. Diagnosis results were obtained for different designs of the SVMs, e.g., for various kernel functions. Moreover, diagnosis results were obtained to highlight the comparison between traditional artificial neural networks and the proposed multi-kernel SVMs. Incremental SVM-based diagnosis results were next compared with nonincremental SVM-based methods. Diagnosis results show that the training time is reduced significantly if we implement incremental learning in linear-kernel SVMs. Furthermore, iMK-SVM-based diagnosis results show high classification rates but low training time in each epoch, thereby providing a practical method for designing an adaptive diagnostic system.

In order to assess the performance of the classifier and its ability to accurately predict the root cause of a failure on a new board, we use a *cross-validation* method to partition the training set into $k$ groups, namely $k$-folder cross-validation [13]. Each group is regarded as a test case while all of the other cases are fed for training. In our work, we assess our model by using a special type of cross-validation method, namely *leave-one-out* (LOO), where the number of partitions $k$ is the same as the total number of cases. For example, for Board 2, each board instance is iteratively selected to be the test case. Classification models are based on the remaining 155 training cases. In LOO estimation, the total number of cases in the testing set is same as the total number of available successfully repaired boards.

To ensure real-time diagnosis and repair, we assume that we are allowed at most three attempts to replace the potential failing components. Success ratio (SR) is the ratio of the number of correctly diagnosed cases to the total number of cases in the testing set. We define $SR_1$ as the success ratio corresponding to the case that the board is deemed to have been successfully repaired only when the actual faulty component is identified and placed at the top of the list of candidates. We also define $SR_2$ ($SR_3$) as the success ratio corresponding to the case that a board is deemed to have been successfully repaired if the actual faulty component is in the first two (three) positions in the list of candidates. In the last column in Table 2.2a, we can see that $SR_1$ is 80.2 %. If three attempts are allowed in the repair process, 91.9 % of the boards can be successfully repaired. These results are a significant improvement over other approaches reported recently in the literature [1, 2]. The SR values for Board 2 are lower (Table 2.2b). Nevertheless, tangible improvement is obtained over other methods, and the diagnosis accuracy is higher than that for methods currently used in production. The training data and evaluation methods will also continue to be used in the following chapters.

**Table 2.2**  Diagnosis results using ANN, SVMs with different kernel functions, and multi-kernel SVMs

|  | ANN | SVM methods | | | | | | | |
|---|---|---|---|---|---|---|---|---|---|
|  |  | Linear kernel | Polynomial kernel | | | Gaussian kernel | | | Multi-kernel |
|  |  |  | $d = 2$ | $d = 3$ | $d = 4$ | $\sigma = 0.5$ | $\sigma = 1$ | $\sigma = 5$ |  |
| (a) Board 1 | | | | | | | | | |
| $SR_1$ | 67.9% | 73.2% | 74.4% | 72.3% | 74.9% | 62.6% | 65.1% | 62.6% | 80.2% |
| $SR_2$ | 78.1% | 80.4% | 82.2% | 81.2% | 82.7% | 74.0% | 76.3% | 74.5% | 85.4% |
| $SR_3$ | 84.4% | 88.2% | 91.9% | 90.4% | 91.9% | 79.2% | 82.5% | 79.3% | 91.9% |
| Training time (s) | 71.2 | 43.2 | 45.2 | 41.1 | 42.0 | 49.9 | 50.3 | 50.1 | 5963 |
| (b) Board 2 | | | | | | | | | |
| $SR_1$ | 57.9% | 66.3% | 63.2% | 63.2% | 66.3% | 66.3% | 60.3% | 55.4% | 71.4% |
| $SR_2$ | 70.3% | 74.3% | 70.1% | 70.1% | 75.5% | 74.3% | 67.8% | 67.8% | 77.8% |
| $SR_3$ | 75.1% | 84.1% | 79.5% | 78.5% | 82.1% | 83.5% | 72.6% | 70.2% | 83.7% |
| Training time (s) | 60.2 | 23.6 | 25.9 | 21.6 | 22.5 | 29.1 | 35.4 | 36.3 | 3891 |

## 2.4.1  Evaluation of MK-SVM-Based Diagnosis System

We use a combination of seven kernels in the multi-kernel machine, including linear kernel, Gaussian kernel ($\sigma$ values of 0.5, 1, and 5), and polynomial kernel (degree $d = 2, 3, 4$). Diagnosis results are shown in Table 2.2a for Board 1 and Table 2.2b for Board 2. For Board 1, the $SR_1$ for the multi-kernel SVM is 6–18 % higher than for the single-kernel SVMs and the ANNs. When the $SR_3$ is considered, the performance of multi-kernel SVMs is similar to that for the single-kernel SVMs. For Board 2, similar improvement of diagnosis accuracy by using multi-kernel SVM is obtained in Table 2.2b. The use of multi-kernel SVM technology leads to a considerable improvement in diagnosis accuracy, but the training time of the multi-kernel SVMs is higher compared to single-kernel SVMs and the ANN. For example, training MK-SVM for Board 1 requires up to an hour as compared to only tens of seconds using single-kernel SVMs and ANNs. Since the training time depends on the number of root causes, number of syndromes, number of cases, and number of iterations required for convergence as described in Sect. 2.3, the training time of multi-kernel SVMs increases quadratically with the board complexity and the number of failing boards that are returned for repair.

### 2.4.2   Evaluation of Incremental SVM-Based Diagnosis System

We implemented linear SVM training on Board 1 and Board 2. The results of incremental learning for linear-kernel SVMs are shown in Figs. 2.6 and 2.8 for Board 1, and in Figs. 2.7 and 2.9 for Board 2. For example, we have a total of 813 training cases for Board 1 as described in Table 2.1. Initially, 800 training cases are randomly selected to build a base SVM model. In the second epoch, 100 more new training cases are randomly selected from the remaining pool. The existing SVM model is updated by using incremental learning to append these new training cases. A total of 100 more new training cases are added into SVM models in the next epoch and so forth. We also construct a nonincremental learning SVM from scratch in each epoch with the same number of training cases as in the corresponding incremental learning SVMs. Figure 2.6 shows that the training time for nonincremental learning SVMs increases linearly with the number of training cases, but the training time for incremental learning remains nearly constant in each epoch, even though the number of training cases increases. In the last epoch when 100 more training cases are appended to the existing SVM with 1500 training cases, the training time of SVMs using incremental learning is 8.27 s compared to 43.2 s using nonincremental learning. This quantifiable reduction in training time using incremental learning can be a significant benefit if thousands of failing boards are returned for repair in high-volume manufacturing. Incremental learning also reduces computational complexity and memory required for training, as described in Sect. 2.3. The use of incremental learning can help reduce the computational complexity of SVMs as described in Sect. 2.3.

**Fig. 2.6** Comparison of training time in each epoch between incremental and nonincremental linear-kernel SVMs for Board 1

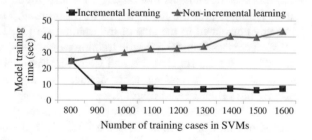

**Fig. 2.7** Comparison of training time in each epoch between incremental and nonincremental linear-kernel SVMs for Board 2

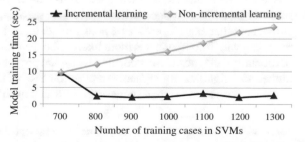

**Fig. 2.8** Comparison of $SR_1$ between incremental and nonincremental linear-kernel SVMs for Board 1

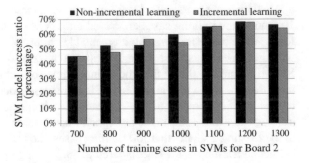

**Fig. 2.9** Comparison of $SR_1$ between incremental and nonincremental linear-kernel SVMs for Board 2

The comparisons of $SR_1$ between incremental learning and nonincremental learning are shown in Figs. 2.8 and 2.9. We found that there is little difference in $SR_1$ between incremental learning and nonincremental learning, thus nonincremental learning can be replaced by incremental learning with the similar success ratios, but the training time is reduced using incremental learning. The observations on $SR_2$ and $SR_3$ are the same as that on $SR_1$. We also found that the $SR_1$ increases when more failing boards are fed to the SVM for training. The diagnosis success ratio is strongly correlated to the size of the training set [1]. The $SR_1$ of the linear-kernel SVMs increases from 59.5 % with training cases of 800 to 73.2 % with training cases of 1600 in Fig. 2.8 for Board 1, and from 45.2 % in Table 2.2b to 66.3 % in Fig. 2.9 for Board 2. The increasing trends of success ratios can also be found in $SR_2$ and $SR_3$ for both Board 1 and Board 2.

In order to evaluate our results in a practical context, we consider what is typically carried out in a board manufacturing line. Most diagnosis and repair actions still rely on the technician's experience and trial-and-error methods. Current diagnostic software used in the production line from where we obtained the boards considers any component that exhibits error as a fault candidate, and no suggestions are provided regarding which component is more likely to be the root cause. Compared to the $SR_1$ of the current diagnostic method, the $SR_1$ for the proposed method is about two times higher, and significantly higher than even the $SR_3$ of the currently used diagnostic method.[1] Based on the repair suggestions provided by the currently deployed method,

---

[1]Exact success ratio for the deployed system are not presented here in order to protect company confidential data.

the debug time for this particular functional test is as high as several weeks, which is clearly not feasible in practice. Debug efficiency is therefore expected to be improved considerably with the accurate repair suggestions provided by the proposed method.

### 2.4.3 Evaluation of Incremental MK-SVM-Based Diagnosis System

Incremental learning can also be applied to MK-SVM training. The success ratio results are shown in Fig. 2.10 for Board 1 and in Fig. 2.11 for Board 2. When we increase the size of the training set, the success ratios of up to three attempts for both Board 1 and Board 2 increase. This observation supports the positive correlation found in [1] between the number of failing boards available for training and the diagnosis accuracy on new boards. The $SR_1$, $SR_2$, and $SR_3$ are 80.2, 85.4, and 91.9 %, respectively, when 1600 training cases are used in the iMK-SVM model for Board 1, and 71.4, 77.8, and 83.7 %, respectively, when 1300 training cases are used in the iMK-SVM model for Board 2.

The training time results are shown in Fig. 2.12 for Board 1 and in Fig. 2.13 for Board 2. The training time of incremental learning MK-SVMs in each epoch is much smaller than that for nonincremental learning MK-SVMs. The training time varies in each epoch for Board 1 due to the size of the training data and iterations needed for convergence, as described in Sect. 2.3. Due to the reduction of support vectors by

**Fig. 2.10** Success ratio of using incremental multi-kernel SVMs for Board 1

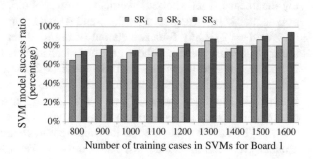

**Fig. 2.11** Success ratio of using incremental multi-kernel SVMs for Board 2

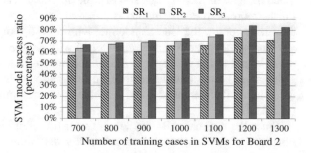

**Fig. 2.12** Comparison of training time in each epoch between using incremental and nonincremental multi-kernel SVMs for Board 1

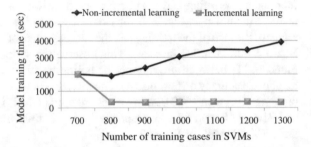

**Fig. 2.13** Comparison of training time in each epoch between using incremental and nonincremental multi-kernel SVMs for Board 2

using incremental learning techniques, the effect of iterations on the total training time can also be reduced in iMK-SVMs, as shown in Fig. 2.12.

The weights in iMK-SVMs training change in each epoch. For example, in Fig. 2.14, we use a combination of 13 kernels in a multi-kernel machine, including the linear kernel, Gaussian kernel ($\sigma$ values of 0.5, 1, 2, 5, 10, 15, 20, and 50), polynomial kernel (degree $d = 2$, 3, and 4), and homogenous polynomial kernel (degree $d = 2$). Only four kernels out of the total set of 14 kernels contribute to the multi-kernel machine; these are the linear kernel, Gaussian kernel with $\sigma = 2$, homogenous polynomial kernel (degree $d = 2$), and polynomial kernel (degree $d = 2$). The weights of the remaining ten kernels are reduced to 0 in the optimized solution. In Fig. 2.14, the weight of the Gaussian kernel with $\sigma = 2$ is 12 % and the weight of homogeneous polynomial kernel with degree $=2$ is 22 % in the first epoch. When more cases are fed for training, the weights of these two kernels are gradually reduced to 0. In the last epoch, only two kernels are left in the multi-kernel machine. And the dominating kernel is the polynomial kernel with degree $=2$ (61 %). Furthermore, kernel distribution is different for different board types. For Board 2, the linear kernel, the Gaussian kernel with $\sigma = 5$, polynomial kernel with degree $=3$, and homogeneous polynomial kernel with degree $=2$ equally contribute to the multi-kernel in the first epoch. After three epochs, the weight of the homogeneous polynomial kernel is reduced to 0. In the final epoch, when 1300 cases are used for training, the weight of linear kernel is 65 % and dominates the diagnosis system (Fig. 2.15).

Since the optimal classifier solutions for different boards lead to different combinations of kernels, we cannot arbitrarily determine a best single kernel for all

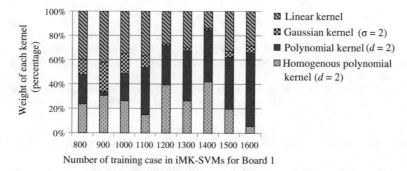

**Fig. 2.14** Illustration of the change in kernel weights in incremental multi-kernel SVMs for Board 1

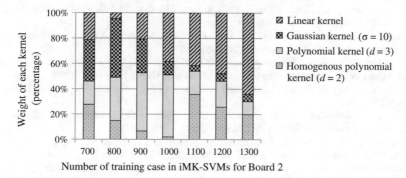

**Fig. 2.15** Illustration of the change in kernel weights in incremental multi-kernel SVMs for Board 2

the boards. The use of iMK-SVMs can adaptively choose the most suitable kernel portfolio for different boards.

## 2.5 Chapter Summary

This chapter has presented a smart diagnosis system based on multi-kernel support vector machines and incremental learning to locate the root cause of functional failures on modern circuit boards. The proposed multi-kernel SVMs method can generate an optimal kernel portfolio to achieve high diagnosis accuracy for board-level functional tests. The use of incremental learning allows the system to adaptively tune the kernel portfolio to achieve high diagnosis accuracy. System training time can also be reduced significantly using incremental learning. Two industrial boards, which are currently in high-volume production, and additional synthetic boards have been used to validate the effectiveness of the diagnosis method. Compared to baseline ANN and several single-kernel SVMs, multi-kernel SVMs show a considerable improvement in diagnostic accuracy based on functional patterns for a real application.

# References

1. Zhang Z, Gu X, Xie Y, Wang Z, Chakrabarty K (2012) Diagnostic system based on support-vector machines for board-level functional diagnosis. In: Proceedings IEEE European test symposium (ETS), pp 1–6
2. Zhang Z, Chakrabarty K, Wang Z, Wang Z, Gu X (2011) Smart diagnosis: efficient board-level diagnosis and repair using artificial neural networks. In: Proceedings IEEE international test conference (ITC), pp 1–10
3. O'Farrill C, Moakil-Chbany M, Eklow B (2005) Optimized reasoning-based diagnosis for non-random, board-level, production defects. In: Proceedings IEEE international test conference (ITC), pp 173–179
4. Zhang Z, Wang Z, Gu X, Chakrabarty K (2010) Board-level fault diagnosis using Bayesian inference. In: Proceedings IEEE VLSI test symposium (VTS), pp 1–6
5. Vapnik V (1995) The nature of statistical learning theory. Springer, Berlin
6. Rakotomamonjy A, Canu S (2008) SVM and Kernel Methods Matlab Toolbox. http://asi.insa-rouen.fr/enseignants/~arakoto/toolbox/index.html
7. Lanckriet G, De Bie T, Cristianini N, Jordan M, Noble W (2004) A statistical framework for genomic data fusion. Bioinformatics 20:2626–2635
8. Varma M, Babu BR (2009) More generality in efficient multiple kernel learning. In: Proceedings of ACM international conference on machine learning (ICML), pp 1065–1072
9. Rakotomamonjy A, Grandvalet Y, Bach F, Canu S (2008) SimpleMKL. J Mach Learn Res 9:2491–2521
10. Chang T, Liu H, Zhou S (2009) Large scale classification with local diversity AdaBoost SVM algorithm. J Syst Eng Electron 20(6):1344–1350
11. Incremental SVM Learning with multiclass support and probabilistic output (2013) http://www-ti.informatik.uni-tuebingen.de/spueler/mcpIncSVM/
12. Neural Network Toolbox (2012) http://www.mathworks.com/products/neuralnet/
13. McLachlan G, Do K, Ambroise C (2004) Analyzing microarray gene expression data, vol 422. Wiley, Hoboken

# Chapter 3
# Diagnosis Using Multiple Classifiers and Majority-Weighted Voting (WMV)

In this chapter, we integrate a meta-learning technique, namely weighted-majority voting (WMV), in our diagnosis system. Multiple classifiers can be leveraged in a WMV-based diagnosis system to incorporate different sets of repair suggestions and form a single set of root-cause candidates. We review the usage of artificial neural networks (ANNs). The advantage of ANNs is its interpretation of the relationship between the syndromes and corresponding faulty components. A trained ANN model associates the output to the weighted inputs, which derives an intuition on the contribution of each input to the output. In addition, the advantage of SVMs, as described in Chap. 2, is that the solution provided by SVMs is globally optimal and unique, while ANNs suffer from multiple local minima. Both of these two methods can be rapidly trained and they are scalable to large datasets. The proposed WMV-based system uses weights to combine the repair suggestions provided by each machine in order to identify a single set of recommended repair suggestions. The proposed WMV system can leverage results from both ANNs and SVMs.

The remainder of this chapter is organized as follows. Section 3.1 reviews the background and states the contributions. Section 3.2 reviews the smart diagnosis system based on ANNs. Section 3.3 highlights the pros and cons of ANNs and SVMs used for functional fault diagnosis. Section 3.4 presents the WMV techniques to optimize the decision outputs of diagnosis by leveraging the repair component candidates provided by ANNs and SVMs. Section 3.5 presented experimental results on three industrial boards, including diagnosis accuracy and time cost. The results of ANNs and SVMs are presented separately based on their parameter settings. The results for the WMV technique are then presented to highlight the improvement in diagnosis accuracy. Section 3.6 concludes the chapter.

© Springer International Publishing Switzerland 2017
F. Ye et al., *Knowledge-Driven Board-Level Functional Fault Diagnosis*,
DOI 10.1007/978-3-319-40210-9_3

## 3.1   Background and Chapter Highlights

The diagnostic resolution offered by today's diagnostic tools is limited to chips on the board by using ATE. No repair guidance is provided for memory devices or passive components on the board. Technicians are forced to run debug programs repeatedly and carry out physical probing in many places to identify the root cause of failures, a practice that significantly increases the debug and repair time. Based on past repair records, we have found the debug time for the functional test considered here to be as high as several weeks. The correctness of diagnosis, i.e., the probability of the actual failing component included in the list of suspects, is unacceptably low, and the root cause is seldom exclusively pinpointed.

The proposed intelligent diagnosis based on machine learning algorithms can be used to overcome the difficulties described above and provide accurate diagnostic conclusions. Several machine learning techniques have been used in board-level diagnosis, such as artificial neural network (ANNs), Bayesian Neural Networks (BNNs), and Support-Vector Machines (SVMs) [1–5].

In [1], a Bayesian inference (BI) machine is used to assist fault diagnosis. Bayesian inference offers a powerful probabilistic method for pattern analysis, classification, and decision making under uncertainty. Learning of BI does not depend on the structural test circuity, which eliminates the requirement of building standard BIST features for all devices on a board. Moreover, by learning the behavior of board/system under different faulty scenarios, the inference engine can automatically perform diagnosis without manual effort.

ANNs, which are inspired by the structure/functional aspects of biological neural networks, are used to model complex relationships between inputs and outputs, or to capture the statistical relationship between observed variables. In [6], a three-layer feedforward network is trained with backpropagation and is designed to target multiple faults in small combinational circuits. The test data are derived from a fault truth table, which is constructed by inserting random single stuck-at faults. The effectiveness of the method is validated only with small circuits that consists fewer than ten logic gates. In [7], the diagnosis of telephone exchange line cards using ANNs at British Telecom is described. A three-layer feedforward network, with 77 input neurons and 8 output neurons, was constructed to solve the diagnosis problem in a structural test. The inputs are measurements from the in-circuit test (ICT). The outputs are fault candidates, which are eight categories of the faulty components (e.g., resistors, relays).

The problems targeted above are different from the functional diagnosis problem addressed in this book. In functional diagnosis, fault syndromes are extracted from a failure log, e.g., mismatched interface, error counter, etc. Therefore, fault syndromes are obtained from the results of functional test sequences, instead of measurements from ICT. Moreover, the diagnostic goal is to accurately locate the faulty component on the board rather than the category of the component. For example, hundreds of resistors can be soldered on a board. We obtain very limited diagnostic information, if we only determine that a resistor caused the failure, without pointing out the specific resistor.

In [2], a carefully crafted architecture of ANNs was used especially for board-level diagnosis. It can be rapidly trained and thus it can handle large datasets with thousands of inputs and outputs. It is initialized with the occurrence probability of fault syndromes, which significantly improves the diagnostic accuracy. It can easily interpret the relationship between the fault syndromes and repair action. The use of diagnosis system can raise the diagnosis accuracy up to 70 %.

In [3] and as described in Chap. 2, another fine-grained machine learning technique has been explored for use in board-level diagnosis. The SVM algorithm is based on the statistical learning theory and the Vapnik–Chervonenkis (VC) dimension introduced in [8]. A Support-Vector Machine (SVM) performs classification by constructing an optimal hyperplane that separates the data into two categories. Compared to ANNs , this new supervised learning method has a number of theoretical and computational merits. A significant advantage of SVMs is that the solution to an SVM is globally optimal and unique, while ANNs suffer from multiple local minima. We analyze the performance of SVMs with real manufacturing data.

However, the limitation of SVMs lie in the selection of a proper kernel. In prior work [3], the best design of SVMs is determined by extensive simulation results. The proposed approach in Chap. 2 overcomes the limitations of single-kernel SVMs by exploiting multikernel SVMs and incremental learning (iMK-SVM) to reduce complexity, achieve significantly higher diagnosis accuracy, and perform reasoning adaptively in realtime as new data becomes available. The kernel function in this approach is defined as a linear combination of different kernels.

Moreover, single diagnosis system suffers from the limitation on its biased diagnosis and provides incorrect repair suggestion. In this chapter, we propose a diagnosis system based on weighted-majority voting. The proposed system uses weights to combine the repair suggestions provided by each machine in order to identify a single set of recommended repair suggestions. The proposed WMV-based diagnosis system can leverage results from both ANNs and SVMs.


## 3.2 Artificial Neural Networks (ANN)

Artificial neural networks (ANNs) are widely used for pattern classification and related problems [9]. ANNs consist of neurons and weighted connections between neurons. Neurons are arranged in layers, and weighted connections link the neurons in different layers. A value is associated with each connection, referred to as *weight*, corresponding to the synaptic strength of neuron connections. The behavior of an ANN depends on both the weights and the input-output function, referred to as *transfer function*. This function typically falls into one of three categories, namely, linear, step, and sigmoid. A simple ANN and the computation in an artificial neuron is shown in Fig. 3.1. Two basic network architectures are feedforward and recurrent. In the feedforward architecture, there is no feedback between layers as the network

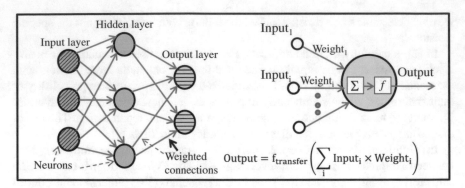

**Fig. 3.1** A simple feedforward neural network and the computation in a neuron [2]

shown in Fig. 3.1. In the recurrent architecture, there is feedback between layers, thus these networks can remember prior inputs. Before use, ANNs must learn from training examples. In supervised learning, we are given a set of example pairs, and the aim is to find an appropriate function that matches the examples. Backpropagation algorithm is widely used for training ANNs, which minimizes the difference between actual outputs of ANNs and the desired values using gradient descent. When the difference is less than the predefined threshold, referred to as *performance goal*, the training is complete. The design and implementation of ANNs are supported by MATLAB neural network toolbox [10]. A multilayer feedforward neural network can be easily implemented in MATLAB.

ANNs are adopted to address the challenges involved in functional diagnosis at the board level. The intention is to develop an automated diagnosis tool that can learn from the historical repair data, involve less human effort, and provide accurate diagnostic guidance. The automated learning characteristic of ANNs fits this problem well. The ANN technique intelligently constructs the connections between fault syndromes and repair actions, without the need for a complete understanding of the complex functionality of a board.

### 3.2.1 Architecture of ANNs

In contrast to existing applications of ANNs to do fault diagnosis, we use a group of single-layer networks to diagnose the root cause of board-level functional failures. The group size is equal to the number of known faulty components that have been replaced in the past (based on the available repair history). The initialization of weights is based on the occurrence probabilities of fault syndromes, instead of random initialization. For each single-layer network, the input neurons represent fault syndromes, and the single-output neuron represents a component. Details on the extraction of fault syndromes and replaced components are presented in the next subsection. The input value is either 1 or 0. A "1" implies that the fault syndrome appears in the log file; otherwise, it is a "0". The desired (ideal) value at the output neuron is either "0" or "1". A "1" means that the component represented by this network is the root cause of failure, and it should be replaced to repair the malfunctioning board; a "0" means that the corresponding component is not the root cause. The actual value at the output is a fraction between 0 and 1, which can be viewed as the probability of the component being the root cause. A value closer to 1 implies that the component represented by the network is more likely to be the root cause and vice versa. The proposed architecture is generic in the sense that it can be used for the functional diagnosis of various types of systems. For a different system, we only need to prepare a new set of training data for the ANNs, which can be easily achieved by updating the scripts used for syndrome extraction. An example of the architecture of ANNs with 500 syndromes and 100 actions is depicted in Fig. 3.2. The training of the neural networks and diagnosis flow are described below (Fig. 3.3).

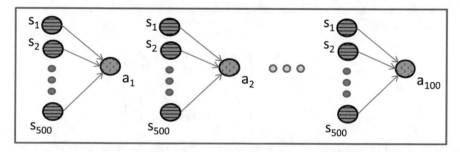

**Fig. 3.2** An illustration of the proposed ANN architecture [2]

**Fig. 3.3** The diagnosis flow using neural networks [2]

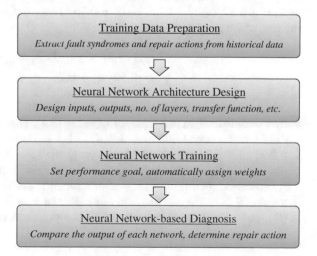

### 3.2.2 Demonstration of ANN-Based Diagnosis System

To illustrate the training and testing procedure of the proposed ANN architecture, consider a hypothetical demonstration board with six cases that are used for training. We built an ANN model to identify faults for new cases. All the test-outcome information is stored in a log file. The extracted syndromes and replaced components are used as inputs and outputs, respectively, for the training of the ANNs. Let $x_1$, $x_2$, and $x_3$ be three syndromes. If the syndrome manifests itself, we record it as 1, and 0 otherwise. Let us suppose that the board has two candidate root causes A and B, and we encode them as $y = [1\ 0]$ and $y = [0\ 1]$, respectively. Note that the dimension of $y$ is as same as the number of candidate root causes. Each $y$ is a vector of only a single 1 and the remaining elements are 0s. In a real scenario, fault syndromes vary across products and tests. Here, we merge the syndromes and the known root causes into one matrix $\mathscr{A} = [\mathscr{B}|\mathscr{C}]$, where the left ($\mathscr{B}$) side refers to syndromes, while the right side ($\mathscr{C}$) refers to the corresponding fault classes. The resulting matrix, shown below in Eq. (3.1), represents the training information for the ANNs.

$$\mathscr{A} = [\mathscr{B}|\mathscr{C}] = \begin{bmatrix} 1 & 1 & 0 & 1 & 0 \\ 1 & 1 & 1 & 1 & 0 \\ 1 & 1 & 0 & 1 & 0 \\ 0 & 1 & 1 & 0 & 1 \\ 0 & 0 & 1 & 0 & 1 \\ 0 & 0 & 1 & 0 & 1 \end{bmatrix} \tag{3.1}$$

We consider the transfer function to be *purelin* (linear transfer function) and the training function to be *trainlm* (LM-optimization). After training this one-layer system of ANNs, we obtain two weights vectors. Fault cause A is represented by the first weight vector $\omega_1 = [\omega_{1-1}\ \omega_{1-2}\ \omega_{1-3}]$, where $\omega_{1-1} = 1.00$, $\omega_{1-2} = 3.26e^{-7}$, $\omega_{1-3} = 8.51e^{-8}$. Fault cause B is represented by the second weight vector $\omega_2$, where $\omega_{2-1} = -0.50$, $\omega_{2-2} = 0.25$, $\omega_{2-3} = 0.75$. Therefore, the vectors for determining the root cause are generated as follows:

$$f(x) = \begin{pmatrix} 1.000 \cdot x_1 + 3.26e^{-7} \cdot x_2 + 8.51e^{-8} \cdot x_3 \\ -0.50 \cdot x_1 + 0.25 \cdot x_2 + 0.75 \cdot x_3 \end{pmatrix} \tag{3.2}$$

Next, suppose a new failing board is received and it has the syndrome [1 1 0], which corresponds to the first row (case) of $\mathscr{A}$ in Eq. (3.1). The function $y$ is evaluated using Eq. (3.2), and since $(1.000 \cdot 1 + 3.26e^{-7} \cdot 1 + 8.51e^{-8} \cdot 0) = 1.00$, while $(-0.50 \cdot 1 + 0.25 \cdot 1 + 0.75 \cdot 0) = -0.25$, so $y = [1\ 0]$. Thus the root cause for this failing board is determined to be A. Suppose a second new failing board with syndrome [0 1 0] in received. In this case, the decision function evaluates to $y = [0\ 1]$; hence, we determine B to be the root cause in this case. For boards with the root cause of class A (B), we can replace the corresponding component A (B).

## 3.3 Comparison Between ANNs and SVMs

The development of ANNs followed a heuristic path, with applications and extensive experimentation preceding theory [2]. In contrast, the development of SVMs involved sound statistical learning theory, implementation, and experiments. The fundamental difference between the two approaches is that ANNs minimize empirical risk (misclassifications in the training set), but SVMs minimize the risk of misclassifying cases of the test set. Therefore, SVMs are less prone to overfitting and often outperform ANNs in practice.

ANNs use a backpropagation learning algorithm to search for the minimum of the error function (mean squared error) in the weight space. The combination of weights, which minimizes the error function, is considered to be a solution of the learning problem. However, the ANN solution obtained using a backpropagation algorithm can only converge to a local minimum. In contrast, the solution to an SVM is global and unique, because SVMs are formulated as a convex quadratic optimization problem.

In addition, the computational complexity of SVMs does not depend on the dimensionality of the input space. SVMs do not attempt to control model complexity by keeping the dimension of the input vectors small. In order to linearly separate the cases, SVMs sometimes transfer the input space to higher dimensional feature space. The model complexity of SVMs is automatically determined in the quadratic programming procedure by selecting the support vectors. One more advantage of SVMs is the simple geometric interpretation, which allows us to easily maximize the margin.

## 3.4 Diagnosis Using Weighted-Majority Voting

Sections 2.2 and 3.2 described diagnosis systems based on two machine learning methods, namely ANNs and SVMs. We regard these two machine learning methods as two *expert systems*. Each system provides its own repair suggestions based on the separate diagnosis techniques. Since we cannot predict which system is more accurate in general, we expect diagnosis with higher resolution to be achieved through a combination of the two systems.

### 3.4.1 Weighted-Majority Voting

Weighted-majority voting (WMV) is a meta-learning reasoning procedure [11]. The first type of majority voting refers to the decision when all experts agree on the same output (unanimous voting). The second type of majority voting refers to the decision when at least one more than half the number of experts agree on the same output (simple majority voting). The third majority voting approach is called weighted-majority voting. If we have evidence that certain experts are more qualified than others, weighting the decisions of those qualified experts more heavily may further improve the overall performance than that obtained by plurality voting. Let us denote the decision of classifier $C_i$ on class $l_j$ as $d_{i,j}$, such that $d_{i,j}$ is 1 if Classifier $C_i$ selects $l_j$ and 0 otherwise. Furthermore, assume that we have a way of estimating the performance of each classifier, and we assign weight $w_i$ to classifier $C_i$ in proportion to its estimated performance. According to this notation, the classifiers whose decision are combined through weighted majority voting will choose class $J$ if

$$\sum_{i=1}^{I} w_i d_{i,J} \geq \max_{j=1}^{C} \left\{ \sum_{i=1}^{I} w_i d_{i,j} \right\} \tag{3.3}$$

i.e., if the total weighted vote received by $l_j$ is greater than or equal to the total vote received by any other class. Weighted-majority voting is illustrated in Fig. 3.4.

The weight $w_i$ is associated with the estimated performance of classifier $C_i$. If we can predict which classifier is more efficient, we can assign the highest weight to this classifier, or even use this classifier alone. In the absence of this knowledge, a plausible strategy is to use the performance of a classifier on a separate validation dataset, or even its performance on the training dataset, as an estimate of that classifier's future performance. The weighting strategy is to assign weights as follows:

$$w_i = \log_2(1/\beta_i) \tag{3.4}$$

where $\beta_i = \varepsilon_i/(1 - \varepsilon_i)$, and $\varepsilon_i$ is the weighted training error of Classifier $C_i$. The training error $\varepsilon_i$ indicates the fitting degree of Classifier $C_i$ to the training data. This estimation has been validated in [11]. For easier interpretation, we can normalize these weights so that they sum up to 1; however, normalization does not change the outcome of weighted majority voting.

**Fig. 3.4** Illustration of weighted-majority voting

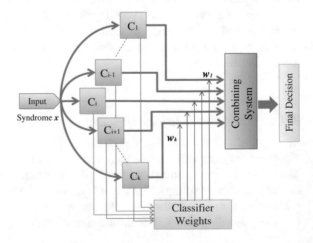

### 3.4.2 Demonstration of WMV-Based Diagnosis System

Consider the top-three repair suggestions provided by SVMs and ANNs. Suppose the repair suggestions provided by SVMs are $s_1$, $s_2$, and $s_3$, and that by ANNs are $a_1$, $a_2$, and $a_3$. The classifier weights $w_{svm}$ and $w_{ann}$ can be immediately calculated based on their empirical training errors. We also assign parameters $m_1$, $m_2$, and $m_3$ (refer to the WMV vector weights) to the corresponding ranks of the three repair suggestions.

Consider a hypothetical demonstration board and suppose that the board has three candidate root causes A, B, and C. After the SVM and ANN classifiers are trained on the same training set, suppose we have $\varepsilon_{svm} = 0.07$ and $\varepsilon_{ann} = 0.10$. Therefore, we can calculate the weights using Eq. (3.4). We obtain the classifier weights $w_{svm} = 3.73$ and $w_{ann} = 3.10$. We also arbitrarily define $m_1 = 4, m_2 = 2$, and $m_3 = 1$ for demonstration purposes.

Suppose we use SVMs to get the repair suggestions A, B, and C in descending priority order. The repair suggestions provided by ANNs are ordered as A, C, and B. We have an intuition of the final decision rank that the root cause A should be the first attempt because both classifiers reach the same root cause. Root cause B should be the second attempt since SVMs are more reliable than ANNs due to the lower training error. Finally, root cause C should be considered as the third attempt.

We next apply WMV on the repair suggestions provided by SVMs and ANNs. According to Eq. (3.3), the majority-voting vector [A, B, C] is $[(m_1 \cdot w_{svm} + m_1 \cdot w_{ann}) \ (m_2 \cdot w_{svm} + m_3 \cdot w_{ann}) \ (m_3 \cdot w_{svm} + m_2 \cdot w_{ann})] = [27.3, 10.6, 9.9]$. This rank is also A, B, and C, which supports our intuition.

Consider another example with four candidate root causes A, B, C, and D. Suppose SVM-based diagnosis provides repair suggestions A, B, and C in descending priority order, but the actual root cause is C. Without additional classifiers, two incorrect repair actions will be taken before the board is successfully repaired. Next, we use WMV with additional input from the ANN model that provides a ranked order of candidates C, B, and D. Suppose we have the same WMV vector weights as in the previous example, and the ANN classifier is weaker than the SVM classifier, where $\varepsilon_{svm} = 0.07$ and $\varepsilon_{ann} = 0.10$. Therefore, we obtain the same classifier weights as above, i.e., $w_{svm} = 3.73$ and $w_{ann} = 3.10$. Now, the WMV vector [A, B, C, D] is [14.9, 13.8, 16.41, 3.17]. In this case, the root cause candidates are re-ranked to be C, A, B, and D. The candidate root cause C is raised from the third attempt to the first attempt, and repair can therefore be successfully carried out with the first attempt.

## 3.5 Results

Experiments were performed on three industrial boards that are currently in high-volume production. Relevant information about two boards is provided in Table 2.1, and reproduced in Table 3.1. For the third board, a total of 1023 repaired boards are

**Table 3.1** Information about the industrial boards used for classification and the log data available

|                                          | Board 1 | Board 2 | Board 3 |
| ---------------------------------------- | ------- | ------- | ------- |
| Number of syndromes                      | 546     | 375     | 751     |
| Number of repair candidates (components) | 153     | 116     | 109     |
| Number of boards                         | 1613    | 1351    | 1023    |

analyzed as training data. A total of 751 fault syndromes are extracted from failure logs. The number of faulty components for repair action is 109. Information about three boards are provided in Table 3.1.

ANNs are implemented using the Neural Network Toolkit [10]. SVMs are implemented using an open source SVM toolbox [12]. Diagnosis results were obtained for different designs of the ANNs, e.g., for various training functions, transfer functions, and number of neuron layers. The proposed one-layer ANNs with linear transfer function and LM-optimization training function has higher diagnosis accuracy and resolution. Diagnosis results for SVMs also vary with different parameters, e.g., kernel functions, kernel degrees, and penalty parameters. SVMs of linear kernel outperform (in terms of diagnosis accuracy) over the ANNs and the SVMs of other kernels. Furthermore, the diagnosis system based on weighted-majority voting provides even higher classification rates, thereby achieving better performance than the single machines. Experiments were run on a 64-bit Linux system with 12 GB of RAM and quadcore Intel i7 processors running at 2.67 GHz.

In this work, we use the same *cross-validation* methods and concepts of success ratios, as described in Chap. 2, to evaluate the performance of ANN-based, SVM-based, and WMV-based diagnosis systems.

### 3.5.1  Evaluation of ANNs-Based Diagnosis System

The diagnostic accuracy of the ANNs-based method is related to the transfer function and training algorithm. In Table 3.2, we present the results of two training functions (*trainrp* and *trainlm*), and two transfer functions (linear and sigmoid). Function *trainrp* updates weight and bias values according to the resilient backpropagation algorithm. Function *trainlm* updates weight and bias values according to LM optimization.

From the Table 3.2, we observe that the best combination of transfer function and training function in terms of diagnostic accuracy and training time for board diagnosis is *trainlm* and linear, and the corresponding results are the last column of each table. It is fast and often the first-choice ANN algorithm in the MATLAB toolbox [10]. For example in Board 1, the $SR_1$ for ANN of *trainlm* and linear is 71.0%, while the SR is 67.9% for ANN of *trainrp* and linear, 55.0% for ANN of *trainrp* and logsig,

**Table 3.2** Diagnosis results for the ANN-based system

| Transfer function | Logsig | | Linear | |
|---|---|---|---|---|
| Training function | *trainrp* | *trainlm* | *trainrp* | *trainlm* |
| (a) Board 1 | | | | |
| $SR_1$ (%) | 55.0 | 34.3 | 67.9 | 71.0 |
| $SR_2$ (%) | 60.0 | 42.0 | 78.1 | 84.0 |
| $SR_3s$ (%) | 64.0 | 44.0 | 84.4 | 90.6 |
| Training time (s) | 52.8 | 83.1 | 71.2 | 52.2 |
| (b) Board 2 | | | | |
| $SR_1$ (%) | 54.9 | 33.4 | 57.9 | 54.9 |
| $SR_2$ (%) | 66.3 | 33.4 | 70.3 | 70.3 |
| $SR_3$ (%) | 71.2 | 39.6 | 75.1 | 74.5 |
| Training time (s) | 42.4 | 72.3 | 60.2 | 33.1 |
| (c) Board 3 | | | | |
| $SR_1$ (%) | 68.7 | 58.0 | 71.3 | 69.4 |
| $SR_2$ (%) | 71.3 | 58.3 | 75.2 | 74.6 |
| $SR_3$ (%) | 72.6 | 58.6 | 76.5 | 76.2 |
| Training time (s) | 142.6 | 420.4 | 217.4 | 96.2 |

and 34.3 % for ANN of *trainrp* and linear. The training time for ANN of *trainlm* and linear is also lower than the other three combinations. For Board 1 in Table 3.2a, the training time for ANN of *trainlm* and linear is 52.2 s, which is lower than the other three where the training times are 52.8, 83.1, and 71.2 s, respectively.

We also found that the diagnosis accuracy is related to the number of training cases. Board 1 and Board 3 have more training cases than Board 2. The classification models of these two boards can be trained more accurately; thus the diagnosis accuracies of these two boards are higher. We observe that the training time is correlated to the number of classes, since the number of constructed ANN networks is as same as the number of classes, as described in Scct. 3.2. For Board 3, 109 neural networks are constructed for 109 root fault classes, which is much more than that for Board 1 and Board 2.

We compare the diagnosis success ratio obtained using the proposed ANNs to that obtained using traditional ANNs and Bayesian inference (BI) [1], with the same training cases and test cases. For traditional ANNs, we consider a three-layer network [7]. The number of neurons in hidden layer is 20. We adjust the number of neurons in the hidden layer to observe the variation of the $SR_1$. A total of six different numbers are used, respectively, 5, 10, 20, 50, 100, and 500. For example, for Board 1 in Table 3.3a, when the number of neurons in the hidden layer is equal to 20, we obtain the highest $SR_1$ to be (60.3 %). When the number is equal to 500, we obtain the lowest $SR_1$ to be (11.4 %). In the comparison shown in Table 3.3, we list the highest $SR_1$ achieved by traditional ANNs. However, it is still much lower than the success ratio achieved by the proposed ANNs. The difference of 16 % in the $SR_1$ and

**Table 3.3** Comparison of SRs obtained by ANNs and BI

|  | $SR_1$ (%) | $SR_2$ (%) | $SR_3$ (%) | Training time |
|---|---|---|---|---|
| (a) Board 1 | | | | |
| Proposed ANNs | 67.9 | 78.1 | 84.4 | 71.3 s |
| One-layer ANNs (no weight initialization) | 66.7 | 78.1 | 82.0 | 72.8 s |
| Traditional ANNs (20 hidden neurons) | 60.3 | 70.7 | 75.3 | 4 h |
| Bayesian inference | 62.5 | 72.5 | 76.2 | 21.2 s |
| (b) Board 2 | | | | |
| Proposed ANNs | 57.9 | 70.3 | 75.1 | 60.2 s |
| One-layer ANNs (no weight initialization) | 53.3 | 61.6 | 73.0 | 51.6 s |
| Traditional ANNs (20 hidden neurons) | 46.6 | 63.3 | 79.6 | 1 h |
| Bayesian inference | 42.1 | 55.6 | 62.4 | 15.2 s |
| (c) Board 3 | | | | |
| Proposed ANNs | 69.4 | 74.6 | 76.2 | 46.2 s |
| One-layer ANNs (no weight initialization) | 53.2 | 66.5 | 70.8 | 47.9 s |
| Traditional ANNs (20 hidden neurons) | 48.9 | 61.4 | 64.2 | 6 h |
| Bayesian inference | 43.7 | 47.3 | 59.3 | 31.4 s |

the difference of 10 % in the success within three attempts is especially significant for high-volume production. The training algorithm used for the architectures is *trainrp*. In the Bayesian-based method, first, we calculate the prior probability based on the training cases, which is the occurrence probability of a syndrome given a repair action. For example, suppose that a particular repair action has been taken 100 times, and a given syndrome showed up 80 times in these 100 cases. Then the prior probability is 0.8. All other occurrence probabilities of syndromes, given repair actions, can be calculated in the same manner. After obtaining the prior probabilities, we can calculate the posterior probability based on Bayes' formula. We take all the fault syndromes into account in the calculation of the posterior probability. For each action, the times of calculation of the posterior probability is equal to the number of fault syndromes. Therefore, the final probability of an action being correct is obtained using the information reflected by all the fault syndromes. More details are described in [1].

The success ratios within three attempts are listed in Table 3.3. We can see that the proposed ANNs-based method has an obvious advantage over the Bayesian inference in terms of success ratio. For example in Board 1, the $SR_1$ of the proposed ANNs, ANNs without initial weights, traditional ANNs, and Bayesian inference are 67.9, 66.7, 60.3, and 62.5 %, respectively. For Board 1, the training time of the proposed

**Table 3.4** Comparison of SRs obtained by SVMs with different kernels

| | Linear kernel | Polynomial kernel | | | Gaussian kernel |
|---|---|---|---|---|---|
| | | $d = 2$ | $d = 3$ | $d = 4$ | |
| (a) Board 1 | | | | | |
| $SR_1$ (%) | 73.2 | 74.4 | 72.3 | 74.9 | 62.6 |
| $SR_2$ (%) | 80.4 | 82.2 | 81.2 | 82.7 | 74.0 |
| $SR_3$ (%) | 88.2 | 91.9 | 90.4 | 91.9 | 79.2 |
| Training time (s) | 43.2 | 45.2 | 41.1 | 42.0 | 49.9 |
| (b) Board 2 | | | | | |
| $SR_1$ (%) | 66.3 | 63.2 | 63.2 | 66.3 | 66.3 |
| $SR_2$ (%) | 74.3 | 70.1 | 70.1 | 75.5 | 74.3 |
| $SR_3$ (,%) | 84.1 | 79.5 | 78.5 | 82.1 | 83.5 |
| Training time (s) | 23.6 | 25.9 | 21.6 | 22.5 | 29.1 |
| (c) Board 3 | | | | | |
| $SR_1$ (%) | 73.5 | 70.0 | 69.0 | 69.0 | 71.7 |
| $SR_2$ (%) | 82.7 | 80.3 | 80.3 | 80.3 | 82.1 |
| $SR_3$ (%) | 84.0 | 82.6 | 83.6 | 83.6 | 82.1 |
| Training time (s) | 80.4 | 91.4 | 72.3 | 75.3 | 82.2 |

ANNs, ANNs without initial weights, traditional ANNs, and Bayesian inference are 71.3 s, 72.8 s, 4 h, and 21.2 s, respectively. The training time of a 20-layer ANN for Board 3 is over 6 hours while the training times are less than 1 minute for the other three methods. A 20-layer ANN has many more hidden layers and hidden neurons than the proposed 1-layer ANN. To calculate the weights and offset of hidden layers, we require an exponential amount of CPU time. Therefore, the calculation time for the 20-layer ANNs is exponentially larger than simple 1-layer ANNs.

## 3.5.2  Evaluation of SVMs-Based Diagnosis System

We apply different types of kernel functions to the SVM-based learning system. The linear kernel, the polynomial kernel (degree 2, 3, and 4), and the Gaussian kernel, are commonly used in SVM training and are therefore implemented in our experiments. The diagnosis results for the four boards are listed in Table 3.4. We observe that the linear kernel function provides the highest SR on our test cases. For example, for Board 1 in Table 3.4a, the $SR_1$ for SVMs with the linear kernel is up to 73.2 % and the $SR_3$ is up to 88.2 %. In addition, the training times of linear-kernel SVMs and polynomial-kernel SVMs are less than the training time of ANNs in Tables 3.2 and 3.3. The training time of SVMs in this section is referred to the time consumed in selecting support vectors and calculating weights of the support vectors for all the SVMs using the training set.

Another interesting parameter in the SVM algorithm is the penalty C; a large value of C corresponding to the assignment of a higher penalty to misclassification. In this sense, a higher value of C leads to more effective SVM training in terms of less training errors, therefore improving model generality, which in turn results in a higher diagnosis accuracy. However, a larger penalty C also leads to the overfitting of the SVM model. Detailed results on diagnosis accuracy obtained by varying parameter C are presented in [3]. In [3], the penalty parameters are varied from infinity to 0.001. When C is less than 0.1, the $SR_1$ drop significantly. Therefore, the penalty parameter has to be relatively large to ensure that SVMs can adequately represent the training set, and thus providing accurate diagnosis. Here, we use C = 1000 throughout as in [3] for SVM training. Besides the penalty parameter C, Gaussian kernel-based SVMs require an additional parameter, namely the kernel width $\sigma^2$. A very small or very large $\sigma^2$ leads to either overfitting or underfitting [13]. Nevertheless, we provide in Table 3.4 the highest SRs obtained using SVMs with Gaussian kernels and the best value of $\sigma^2$. The implementation of Gaussian kernel-based SVMs is impractical since $\sigma^2$ must be individually determined for different boards using exhaustive search.

### 3.5.3   Evaluation of WMV-Based Diagnosis System

We next present results for WMV, where a combination of ANN and SVM is used to achieve higher diagnosis accuracy. The weights of two classifiers, $w_{svm}$ and $w_{ann}$, are based on the training errors, which are generated dynamically while the models are built. Linear-kernel SVMs and one-layer ANNs are used as the two learning methods in the combined WMV diagnosis system. We observe that the training errors of SVMs for most boards are less than that of ANNs, with Board 2 being the only exception. The corresponding classifier weights are shown in the captions of Table 3.5. Therefore, in most cases, the decisions provided by the SVM diagnosis subsystem has a greater influence on WMV diagnosis, while ANNs serve as an secondary decision engine to verify or contradict the diagnosis decisions obtained from SVMs.

In our experiment, we investigate the contributions to the SR for three attempts from the two constituent diagnosis systems. Intuitively, we should assign the highest weight to the first attempt, next highest weight to the second attempt, and lowest to the third attempt. However, we cannot arbitrarily judge the optimal configuration for the weight combination. We conducted experiments to assess different weight configurations. Since we select three optimal attempts from a total of three to six potential repair attempts provided by two diagnosis systems, the third attempt in single diagnosis system is least likely to be the correct choice compared to the first two attempts. Thus we assign a small weight of 1 to the third attempt. The second attempt is then assigned a larger weight of 2 because it is more accurate than the third attempt. We consider the weights of the first attempt to be 3, 4, and 5. Therefore, the

**Table 3.5** Comparison of SRs obtained by ANNs, SVMs, and WMV

| | $SR_1$ (%) | $SR_2$ (%) | $SR_3$ (%) |
|---|---|---|---|
| (a) Board 1 $w_{svm} = 6.63$ and $w_{ann} = 5.19$) | | | |
| Proposed ANNs | 67.9 | 78.1 | 84.4 |
| Proposed SVMs | 73.2 | 80.4 | 88.2 |
| WMV (3-2-1) | 78.5 | 80.4 | 89.5 |
| WMV (4-2-1) | 78.5 | 82.5 | 89.5 |
| WMV (5-2-1) | 78.5 | 82.5 | 89.5 |
| (b) Board 2 $w_{svm} = 3.63$ and $w_{ann} = 2.19$) | | | |
| Proposed ANNs | 57.9 | 70.3 | 75.1 |
| Proposed SVMs | 66.3 | 74.3 | 84.1 |
| WMV (3-2-1) | 69.0 | 75.0 | 87.0 |
| WMV (4-2-1) | 69.0 | 76.7 | 87.0 |
| WMV (5-2-1) | 69.0 | 76.7 | 89.4 |
| (c) Board 3 ($w_{svm} = 6.63$ and $w_{ann} = 5.19$) | | | |
| Proposed ANNs | 69.4 | 74.6 | 76.3 |
| Proposed SVMs | 73.5 | 82.7 | 84.0 |
| WMV (3-2-1) | 73.0 | 83.4 | 86.8 |
| WMV (4-2-1) | 75.6 | 83.4 | 86.8 |
| WMV (5-2-1) | 75.6 | 83.4 | 86.8 |

weighting strategies are 3-2-1, 4-2-1, and 5-2-1. The choice of weight is empirical, but these weights define the relative importance of different repair attempts.

The diagnosis results are shown in Table 3.5. We observe that the diagnosis accuracy for WMV is at least as good as the best single diagnosis system. In particular, WMV performs better for boards with low volume and where the diagnosis accuracy is low for a single diagnosis system. We observe the improvements on Board 1 in Table 3.5a. The $SR_1$ is raised from 67.9 % using ANNs and 73.2 % using SVMs to 78.5 % using WMV. The SR for third attempt is raised from 84.4 % using ANNs and 88.2 % in SVMs to 89.5 % using WMV. We also observe that the choices of different weights do not show substantial difference in diagnosis accuracy; However, we recommend the weighting strategy of 4-2-1 since this set of weights always provide the highest diagnosis accuracy in this chapter among the three weighting strategies.

We illustrate the improvement in diagnosis accuracy by comparing relative SR percentage increase for the three boards; see Figs. 3.5 and 3.6. The WMV method uses the weights of 4-2-1. We observe that the SR improvements for boards with fewer training cases are more than those with more training cases. Let $SR_i^A$ ($SR_i^S$) be the SR with $i$ attempts for the baseline method ANNs (SVMs). Let $SR_i^W$ be the SR achieved with $i$ attempts using WMV. The relative percentage increase in SRs is defined as $rSR_i^A = ((SR_i^W - SR_i^A)/SR_i^A) \times 100\%$ in Fig. 3.5 and $rSR_i^S = ((SR_i^W - SR_i^S)/SR_i^S) \times 100\%$ in Fig. 3.6. For Board 1, $SR_1$, $SR_2$, and $SR_3$ increase

**Fig. 3.5** SR improvement when WMV is used instead of ANNs

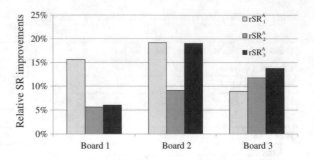

**Fig. 3.6** SR improvement when WMV is used instead of SVMs

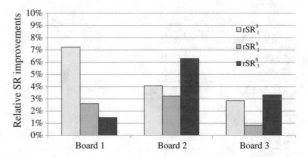

from 67.9, 78.1, and 84.4 % for the ANN baseline and 73.2, 80.4, and 88.2 % for the SVM baseline to 78.5, 82.5, and 89.5 % using WMV. Therefore, the values of $rSR_1^A$, $rSR_2^A$, and $rSR_3^A$ for Board 1 are 15.9, 5.3, and 7.1 %, respectively in Fig. 3.5 and the values of $rSR_1^S$, $rSR_2^S$, and $rSR_3^S$ are 7.3, 2.6, and 1.7 %, respectively in Fig. 3.6.

## 3.6   Chapter Summary

The proposed method is a significant improvement over the traditional manual diagnosis and diagnostic software currently used in production. With three attempts, the diagnostic success ratio of the diagnostic software currently used in the production line is less than 50 %[1] and even poorer when the volume of boards are small. The first attempt SR is considerably lower. Using the proposed diagnostic system, more faulty boards can be successfully repaired within three attempts. To better understand what these results imply, consider the following hypothetical (yet representative) example. Suppose 5000 boards are fabricated per month by a manufacturer; each board costs \$3000 and 10 % of the boards fail functional test. If we can improve the repair success ratio by even a small percentage, for example 5 %, using the new diagnostic system, this means that the board manufacturer saves \$75,000 per month for every 5 % improvements in success ratios.

---

[1]Exact success ratios are not presented here in order to protect company-confidential data.

This chapter has described a smart diagnosis system based on a combination of artificial neural networks and support-vector machines to rapidly and correctly locate the root failure causes on modern circuit boards. The proposed diagnosis system was designed to automate syndrome extraction or preparation for diagnosis and system training. The relationship between log files and potential fault components cannot be inferred from visual inspection. Artificial neural networks can generate a visual relationship between syndrome and root cause, and the support-vector machine creates an optimal hyperplane to separate the root cause in syndrome space. Weighted-majority voting was proposed to take advantage of both ANNs and SVMs to provide an optimal repair suggestion set. Three industrial boards, which are currently in high-volume production were used to validate the effectiveness of the proposed method.

# References

1. Zhang Z, Wang Z, Gu X, Chakrabarty K (2010) "Board-level fault diagnosis using Bayesian inference". In: Proceedings of the IEEE VLSI test symposium (VTS), p 1–6
2. Zhang Z, Chakrabarty K, Wang Z, Wang Z, Gu X (2011) "Smart diagnosis: efficient board-level diagnosis and repair using artificial neural networks". In: Proceedings of the IEEE international test conference (ITC), pp 1–10
3. Zhang Z, Gu X, Xie Y, Wang Z, Chakrabarty K (2012) "Diagnostic system based on support-vector machines for board-level functional diagnosis". In: Proceedings of the IEEE European test symposium (ETS), p 1–6
4. Ye F, Zhang Z, Chakrabarty K, Gu X (2013) Board-level functional fault diagnosis using artificial neural networks, support-vector machines, and weighted-majority voting. IEEE Trans Comput-Aided Des Int Circuits Syst (TCAD) 32(5):723–736
5. Ye F, Zhang Z, Chakrabarty K, Gu X (2013) "Board-level functional fault diagnosis using learning based on incremental support-vector machines". In: Proceedings of the IEEE Asian test symposium (ATS), pp 208–213
6. Al-Jumah AA, Arslan T (1998) Artificial neural network based multiple fault diagnosis in digital circuits. Proceedings of the international symposium on circuits and systems (ISCAS), vol 2, pp 304–307
7. Totton K, Limb P (1991) "Experience in using neural networks for electronic diagnosis". In: Proceedings of the ACM international conference on artificial neural networks, pp 115–118
8. Vapnik V (1995) The nature of statistical learning theory. Springer, Heidelberg
9. Haykin S (2008) Neural Networks and Learning Machines. Prentice Hall, New Jersey
10. Neural Network Toolbox (2012). http://www.mathworks.com/products/neuralnet/
11. Littlestone N, Warmuth M (1994) The weighted majority algorithm. J Inf Comput 108(2):212–261
12. Rakotomamonjy A, Canu S (2008) SVM and Kernel Methods Matlab Toolbox. http://asi.insa-rouen.fr/enseignants/~arakoto/toolbox/index.html
13. Keerthi S, Lin C (2003) Asymptotic behaviors of support vector machines with Gaussian kernel. Neural Comput 15(7):1667–1689

# Chapter 4
# Adaptive Diagnosis Using Decision Trees (DT)

Functional fault diagnosis at board-level is desirable for high-volume production since it improves product yield. However, to ensure diagnosis accuracy and effective board repair, a large number of syndromes must be used. Therefore, the diagnosis cost can be prohibitively high due to the increase in diagnosis time and the complexity of syndrome collection/analysis.

In this chapter, we apply decision trees to the problem of adaptive board-level functional fault diagnosis in high-volume manufacturing. The number of syndromes used for diagnosis is significantly less than the number of syndromes used for a priori training. Despite an order of magnitude or higher reduction in the number of syndromes compared to SVMs and ANNs, the diagnosis accuracy of the proposed method is comparable to that for the baseline methods and considerably higher for the same number of syndromes. Another advantage of using decision trees is that it can also be used to select an effective, but reduced, set of syndromes for use by ANNs or SVMs in a subsequent step. Moreover, to enable the reuse of knowledge from the test-design stage, we use an incremental version of decision trees so as to bridge the knowledge gap between test-design stage and volume production stage.

The remainder of this chapter is organized as follows. Section 4.1 presents the background and describes the work contributions. Section 4.2 describes the diagnosis system based on DTs. The architecture of DTs and advantages of this architecture are presented. Section 4.3 describes the use of incremental learning in DTs. Section 4.4 describes the diagnosis flow using incremental DTs. Section 4.5 presents the results for two industrial boards. The substantial reduction in the number of syndromes highlights the effectiveness of the proposed DT-based diagnostic system. Finally, Sect. 4.6 concludes the chapter.

© Springer International Publishing Switzerland 2017
F. Ye et al., *Knowledge-Driven Board-Level Functional Fault Diagnosis*,
DOI 10.1007/978-3-319-40210-9_4

## 4.1   Background and Chapter Highlights

The proposed diagnostic system is based on incremental decision trees (DTs). A trained DT model consists of a tree-like diagnosis flow, described in detail in Sect. 4.2. The discriminative ability of syndrome is analyzed based on statistical criterion, such as Gini Index or Information Gain [1, 2]. The most discriminative syndrome is selected to be the starting point for diagnosis. The final repair suggestion is available at the leaf node of a DT. The diagnosis time can be significantly reduced using DTs, since only a small number of syndromes are needed for identifying the root cause, instead of all the syndromes that are used in ANNs and SVMs [3, 4]. As shown in Sect. 4.5 for industry boards, only tens of syndromes on average are required for identifying the root cause. Furthermore, an advantage of DT is its interpretation of the relationship between syndromes and corresponding root causes.

Unlike the diagnosis procedures of ANNs and SVMs that use all the syndromes simultaneously for diagnosis, the diagnosis procedure of DTs relies on a multi-stage or sequential approach to the problem. In fact, it resembles the steps followed by an experienced debug technician. When a DT-based diagnosis system receives a faulty board, it starts by determining and analyzing the most discriminative syndrome. Based on the results obtained using this syndrome, the diagnosis system can locate a group of potential root causes and also identify the second-most discriminative syndrome, and so on. After several syndromes are determined in this way and analyzed, the most-likely root cause candidate is suggested for repair. In practice, skilled debug engineers usually follow such a tree-like diagnosis procedure, which is derived from their experience. This knowledge can be recorded in the same form of a decision tree. Moreover, in a typical volume production environment, diagnosis knowledge is collected in a dynamic manner as more faulty boards are detected after testing. Hence, the diagnosis system must also adapt to new knowledge corresponding to various error scenarios. By using incremental decision trees, we bridge the knowledge gap between what debug engineers know at test-design stage and the information gained through testing during volume production; see Fig. 4.1.

The merits of using the proposed adaptive functional fault diagnosis system are:

**Fig. 4.1** Illustration of the benefit of bridging the knowledge gap between debug technicians and an automated learning system

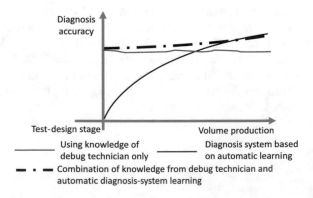

- Reduced and selective set of syndromes needed for diagnosis. Syndrome selection is based on the information derived from the previous syndrome.
- Automatic integration of knowledge of newly identified error scenarios during volume production and field deployment in the field to the DT-based diagnosis system.
- Utilization of knowledge from debug technicians at test-design stage along with the knowledge of failing boards gained continuously in the field.

## 4.2 Decision Trees

Decision trees (DTs) are widely used in statistics, data mining, and machine learning as a predictive model for mapping observations of an item to its targeted value [1]. A final decision is reached based on a series of observations and the associated decision steps. In the early days of decision theory, the DT model was constructed manually. DT learning was subsequently introduced in [2], as a result of which the DT model could be constructed automatically. DTs are now applied to various fields as a powerful learning and classification tool, e.g., market prediction, disease prediction [5], etc.

A DT consists of two types of nodes, leaf(terminal) nodes and decision(internal) nodes. Leaf nodes refer to the nodes that do not branch and contain class information. Decision nodes refer to the nodes that can branch to multiple child nodes or leaf nodes. Based on the inference made from the current decision node, a child node is selected for further branching. Suppose we have a faulty board with several root cause candidates denoted by the elements of the set $A = \{A_1, A_2 \ldots A_m\}$. These root-cause candidates are the leaf nodes in the decision tree. To diagnose the root cause for this faulty board, we have a number of syndromes for diagnosis, i.e., $T = \{T_1, T_2 \ldots T_n\}$, which are encoded in the decision nodes. Each decision node $T_i$ has two branches, i.e., $t_i$ and $\bar{t}_i$, where $t_i$ denotes that the syndrome $T_i$ manifests itself, and $\bar{t}_i$ otherwise. Figure 4.2 illustrates the structure of a decision tree for board diagnosis. For example, two sequences of syndrome observations lead to the prediction of root cause $A_1$ in Fig. 4.2: (1) $t_1 \rightarrow t_2 \rightarrow A_1$; (2) $\bar{t}_1 \rightarrow t_3 \rightarrow t_4 \rightarrow A_1$. Both diagnosis procedures require less syndromes than the total number of syndromes in the tree. A reduction in the number of tests results in significant reduction in diagnosis time.

## 4.2.1 Training of Decision Trees

During the decision-making process, different observations on a current internal node lead to different child nodes for further classification. Therefore, determining the discriminative ability of the feature is critical in decision-tree training. DT training involves the recursive partitioning of the training data, which is split into increasingly homogeneous subsets on the basis of a splitting criterion. There are several common

**Fig. 4.2** Illustration of a
decision tree

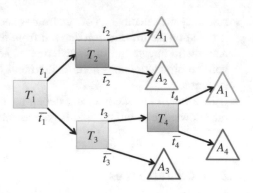

criteria to choose from, such as Information Gain [2], Gini Index [1], and Twoing
[1]. These are described below.

### 4.2.1.1  Information Gain

Let $C$ denote a set of training cases (faulty boards), which can be labeled by a set
of classes (root causes) $A = \{A_1, A_2 \ldots A_m\}$. The test set $T = \{T_1, T_2 \ldots T_n\}$ can be
deemed to be a vector of features attached to the classes, where $n$ is the dimension of
the feature space. Our goal is to select the most discriminative feature $T_i$ from $T$. The
expected information obtained for each feature $T_i$ is measured by its gain $IG(C, T_i)$,
defined below:

$$IG(C, T_i) = E(C) - E(C|T_i) \tag{4.1}$$

where $E(C)$ is the entropy of $C$, i.e., $E(C) = -\sum_j p(A_j) \log p(A_j)$, with $p(A_j)$ being
the probability of case with root cause $A_j \in A$ being in the training set $C$. In the same
way, $E(C|T_i) = -\sum_j p(A_j|t_i) \log p(A_j|\bar{t_i})$, where $t_i$ is the observation of $T_i$ manifests
itself, and $\bar{t_i}$ otherwise.

A rule of thumb is to branch on the feature that provides the most information
gain [2]. information gain.

### 4.2.1.2  Gini Index

The Gini Index criterion is based on Gini impurity $GI(C, T_i)$, given as follows:

$$GI(C, T_i) = Gini(C|T_i) - Gini(C), \tag{4.2}$$

where $Gini(C)$ is the Gini Index of $C$: $Gini(C) = \sum_j p(A_j)(1 - p(A_j))$. In the same
way, $Gini(C|T_i) = \sum_j p(A_j|t_i)(1 - p(A_j|\bar{t_i}))$.

The rule here is to branch on the feature that has the lowest Gini impurity [2].

### 4.2.1.3 Twoing

The Twoing criterion splits classes by grouping the two most discriminative super-classes, rather than identifying each class from the remaining classes as in the Gini Index criterion. This criterion is useful for multiclass classifications. The Twoing value is defined as follows:

$$Two(T_i) = \left( \sum_{k \in m} \left( \frac{L_k}{(T_i)_L} - \frac{R_k}{(T_i)_R} \right) \right)^2, \tag{4.3}$$

where $(T_i)_L / (T_i)_R$ is the number of cases on the left/right of the decision node $T_i$, and $L_k / R_k$ is the number of cases with the super root-cause $A_k$ on the left/right of the decision node $T_i$. The rule here is to branch on the feature that has the lowest Twoing value [2].

Based on the above discussion, we note that DT training generates a tree-like model, where the root node is the most discriminative feature and leaf nodes correspond to the labels of classes. In the prediction stage, the classifier examines the syndromes one-by-one by traversing the branches in the decision tree. In each step, the choice of next syndrome is determined by the observation of the current syndrome. Unlike the diagnosis procedure of ANNs and SVMs that analyze observations of all the syndromes based on a trained classifier, DTs require only a small number of syndromes for diagnosis.

## 4.2.2 Example of DT-Based Training and Diagnosis

We next present an example of training: a DT-based diagnosis system based on a board-repair database. Consider a hypothetical demonstration board with eight cases that are used for training. We build a DT model to identify faults for new cases. All the test-outcome information is stored in a log file, a sample of which is shown in Fig. 1.8. The extracted syndromes and replaced components are used as inputs and outputs for the training of the DTs. Let $T_1, T_2, T_3$, and $T_4$ be four syndromes. If the syndrome $T_i$ manifests itself, we record it as $t_i$, and $\bar{t}_i$ otherwise. In a real scenario, fault syndromes vary across products and tests. Let us suppose that the board has four candidate root-cause components $A_1, A_2, A_3$, and $A_4$, respectively. Here, we merge the syndromes and the known root causes in one matrix $\mathscr{C} = [\mathscr{T} | \mathscr{A}]$, where the left side ($\mathscr{T}$) refers to the outcomes of tests, while the right side ($\mathscr{A}$) refers to the corresponding fault classes. This matrix represents the training information for the DTs.

$$\mathscr{C} = [\mathscr{T}|\mathscr{A}] = \begin{bmatrix} \overline{t_1} & \overline{t_2} & t_3 & t_4 & A_1 \\ \overline{t_1} & \overline{t_2} & \overline{t_3} & t_4 & A_1 \\ \overline{t_1} & \overline{t_2} & t_3 & t_4 & A_1 \\ \overline{t_1} & \overline{t_2} & \overline{t_3} & \overline{t_4} & A_2 \\ \overline{t_1} & t_2 & \overline{t_3} & \overline{t_4} & A_2 \\ t_1 & t_2 & \overline{t_3} & \overline{t_4} & A_3 \\ t_1 & \overline{t_2} & \overline{t_3} & \overline{t_4} & A_3 \\ t_1 & \overline{t_2} & t_3 & t_4 & A_4 \\ t_1 & t_2 & t_3 & t_4 & A_4 \end{bmatrix} \tag{4.4}$$

We consider the splitting criterion based on information gain [2]. We calculate the information gains for all the syndromes and classes. According to Eq. (4.1), information gains for the individual syndromes are calculated to be 1.97 for $T_1$, 1.31 for $T_2$, 1.67 for $T_3$, and 1.97 for $T_4$. Therefore, we select the syndrome $T_1$ with the highest information gain as the root node for our diagnosis system; note that $T_4$ could also have been selected because it provides the same information gain. These nine cases can be divided into two groups, i.e., the first five (collectively referred to as $\mathscr{C}_1$) and the last four (collectively referred to as $\mathscr{C}_2$). We continue to split these two sets of data. For $\mathscr{C}_1$, we can choose $T_4$ with the highest $IG = 0.97$, and $T_4$ is chosen to be the splitting node again for $\mathscr{C}_2$. Since splitting stops when all instances have the same root causes in the subgroup after splitting, the DT in this example consists of three decision nodes and four leaf nodes. Figure 4.3 shows the trained DT. The decision tree consists of one decision node of $T_1$ and two decision nodes arising from $T_4$. For example, the observation sequence $t_1 \rightarrow \overline{t_4}$ gives rise to the prediction of the root cause $A_1$. In this example, the total number of syndromes used for diagnosis is reduced from 4 to 2.

Next, suppose a new failing board is received and it has the syndrome $[\overline{t_1}\ \overline{t_2}\ t_3\ t_4]$, which corresponds to the first row (case) of $\mathscr{C}$ in Eq. (4.4). This faulty board is first evaluated using syndrome $T_1$. Since $\overline{t_1}$ is observed for this board, $T_4$ is then selected for diagnosis. As $t_4$ is obtained for this case, the root cause for this failing board is determined to be $A_1$.

**Fig. 4.3** DT trained for demo board

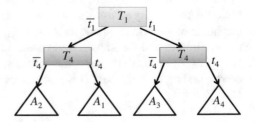

## 4.3 Diagnosis Using Incremental Decision Trees

Diagnosis based on decision trees was used in the past even when DTs were generated manually by debug technicians [6]. Knowledge provided by experienced technicians were incorporated in DTs. However, these DTs cannot handle new error scenarios that become available during volume manufacturing. Technicians must regenerate the complete decision tree to incorporate the new knowledge. Decision trees can be generalized automatically using data analysis and splitting criteria described in the previous section. However, these automatically generated decision trees must also be retrained (regenerated) to deal with the occurrence of new errors and the resulting new knowledge. We tackle this problem using incremental decision trees (iDTs). The DT developed by technicians at test-design time can thus be incrementally and continuously updated to include new errors (Fig. 4.1).

Figure 4.4 illustrates the procedure of training an incremental decision tree: (1) the decision tree is updated by incorporating the knowledge of a case using the procedure *add_case_to_tree*; (2) the tree is checked with the best splitting rule on each decision node recursively using the procedure *ensure_best_splitting*. We demonstrate the use of iDTs using the decision tree trained in Fig. 4.3 based on the past logs. Note that the decision tree can also be derived by an experienced debug technician at the test-design stage.

### 4.3.1 Incremental Tree Node

To enable incremental learning in DTs, a tree node needs to maintain additional information. For each decision node, the counts of cases for each branch, namely *scores*, are also retained. Figure 4.5 shows the decision tree, in which all the decision nodes and leaf nodes are annotated with the corresponding scores. For example, the decision node of $T_1$ contains the score of 5:4, which means five cases have $\overline{t_1}$ derived from $T_1$, while the other four cases have $t_1$. The leaf node corresponding to $A_1$ contains the score of 3, which means a total of three cases with the observation sequence $t_1 \rightarrow \overline{t_4}$ have the root cause $A_1$. Here, we simply use the scores obtained from the board-repair database and Eq. (4.4). Note that if the decision tree is generated by

**Fig. 4.4** Outline of the training algorithm for iDTs

| Function 1: Incremental Decision-Tree Training |
| --- |
| **Input:** root node $T_r$, case $c$, |
| **Output:** root node $T_r$ |
| 1 **Function** *incremental_update* $(T_r, c)$ |
| 2     *add_case_to_tree* $(T_r, c)$ |
| 3     *ensure_best_splitting* $(T_r)$ |
| 4 **End Function** |

technicians, the scores stored in the tree nodes shall be assigned manually depending
on the level of confidence of the technician.

### 4.3.2   Addition of a Case

Once a new case is available, the decision tree is updated by incorporating this case.
The case is considered for inclusion in as many places as possible in the tree according
to the branch rules in the decision nodes. The procedure in Fig. 4.6 demonstrates the
addition of a case in a tree. Three scenarios may occur during the addition of a new
case.

In Scenario I (Fig. 4.7), if the case has the same class label as the leaf node, it is
simply added to the score stored in the leaf node.

For example, suppose we update the tree (Fig. 4.5) using a case with the syndrome
$[\overline{t_1}\ \overline{t_2}\ t_3\ t_4]$ and the root cause $A_1$. The scores in the root node $T_1$, the decision node
$T_4$ (left), and the leaf node $A_1$ are updated, which are highlighted in the figure.

In Scenario II (Fig. 4.8), if the new case has a different class label from the leaf
node, the leaf node is converted to a decision node. All the cases in the previous
leaf node are reorganized. They are transferred to the appropriate (new) leaf nodes
according to the splitting rule on the new decision node. However, it is possible that
one or more values of the cases can be *missing* during tree construction. To account
for this possibility, a missing value is treated as a special value that does not satisfy
the splitting rule at a decision node. In this book, a missing value for a syndrome $T_i$
is treated as $\overline{t_i}$.

For example, suppose we update the existing tree (Fig. 4.7) using a new case with
the syndrome $[\overline{t_1}\ \overline{t_2}\ t_3\ t_4]$ and the root cause $A_4$. Following the tree path $\overline{t_1} \rightarrow t_4 \rightarrow A_1$,
the leaf node of $A_1$ is reached. Here, the new case with the root cause $A_4$ has a different
class label from the leaf node of $A_1$. Hence, this leaf node shall be converted to a
decision tree. Based on the *missing*-value rule, the values of $T_2$ and $T_3$ of the existing
four cases in the prior leaf node are treated as $\overline{t_2}$ and $\overline{t_3}$, respectively. In the new case,
$\overline{t_2}$ and $t_3$ are observed. Based on the information-gain criteria described in Sect. 4.2.1,

**Fig. 4.5** Illustration of
nodes with additional scores
for incremental learning

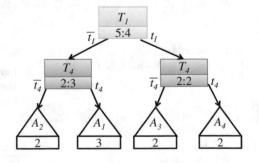

**Fig. 4.6** Procedure for
adding a new case to a DT

| Function 2: Add a case to the tree |
|---|
| **Input:** node $T_i$, case $c$ |
| **Output:** node $T_i$ |
| 1  **Function** *add_case_to_tree*($T_i$, c) |
| 2  **If** $T_i$ is NULL: |
| 3      *convert_to_leaf_node* ($T_i$) |
| 4  **End If** |
| 5  **If** $T_i$ is leaf node: |
| 6      **If** $c$ is same as other cases in $T_i$: |
| 7          Save $c$ in $T_i$ |
| 8      **else If** *is_convertible*($T_i$): |
| 9          *convert_to_internal_node*($T_i$) |
| 10         Save $c$ in $T_i$ |
| 11         Mark $T_i$ as *not stale* |
| 12         $Rl$ = determine_splitting_rule($T_i$) |
| 13         **For** each case $c_j$ in $T_i$: |
| 14             **If** *test*($Rl$, $c_j$) is true: |
| 15                 *add_case_to_tree*($T_i \rightarrow$left, $c_j$) |
| 16             **else:** |
| 17                 *add_case_to_tree*($T_i \rightarrow$right, $c_j$) |
| 18             **End if** |
| 19         **End For** |
| 20     **else:** |
| 21         Mark $T_i$ as *impurity* |
| 22     **else:** |
| 23         Save $c$ in $T_i$; |
| 24         Mark $T_i$ as *stale*; |
| 25         **If** *test*($Rl$, $c$) is true: |
| 26             *add_case_to_tree*($T_i \rightarrow$left, $c$); |
| 27         **else:** |
| 28             *add_case_to_tree*($T_i \rightarrow$right, $c$); |
| 29         **End if** |
| 30     **End if** |
| 31     return $T_i$; |
| 32 **End Function** |

individual information gains are calculated to be 0 for $T_2$, and 0.72 for $T_3$. Therefore, $T_3$ is selected to be the new decision node. The new tree is shown in Fig. 4.8.

In Scenario III, if the new case has a different class label from the leaf node and the leaf node cannot be converted to a decision node, we keep the node as a leaf and marked it as an impurity leaf. The case will still be added to the score on the leaf node, but the case is marked with a different class label. The class label of the

**Fig. 4.7** Illustration of the addition of a case (Scenario I)

**Fig. 4.8** Illustration of the addition of a case (Scenario II)

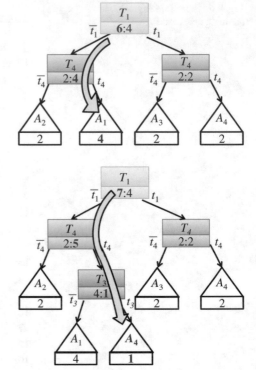

leaf node is determined by the majority class label with the most case-counts on the leaf node. For example, suppose we update the existing tree (Fig. 4.8) using a new case $[\bar{t_1} \, \bar{t_2} \, \bar{t_3} \, t_4]$ and the root cause $A_3$. Following the tree path $\bar{t_1} \rightarrow t_4 \rightarrow \bar{t_3} \rightarrow A_1$, the leaf node has a different class label from the actual root cause of the case. The incremental learning algorithm attempts to turn the leaf node to a decision node. Based on the missing-value rule, the values of $T_2$ of the existing cases are treated as $\bar{t_2}$, which is the same as the observation of $T_2$ on the new case. Thus, $T_2$ cannot be turned to a new decision node, while the remaining three syndromes have already served as the decision nodes. Therefore, the leaf node is marked as an impurity leaf. Since the number of cases with the class label $A_1$ is more than the rest of the class labels on this leaf node, this leaf node is still labeled with $A_1$.

### 4.3.3  Ensuring the Best Splitting

The addition of new cases results in the expansion of the decision tree. The set of cases that satisfied a previous splitting rule is not necessarily admissible for a new rule. Therefore, the tree must be checked with the best splitting rule on each decision node. One way to ensure that each decision node is using the best splitting rule is to

**Fig. 4.9** Pseudo-code for
the algorithm
ensure_best_splitting()

| Function 3: Ensure the best splitting rule |
| --- |
| **Input:** root node $T_r$, case $c$, |
| **Output:** root node $T_r$ |
| 1  **Function** *ensure_best_splitting* $(T_i)$ |
| 2      **If** $T_i$ is a decision node and $T_i$ is *stale*: |
| 3          $Rl$ = determine_splitting_rule$(T_i)$ |
| 4          **If** $Rl$ is current splitting rule: |
| 5              *transpose_tree*$(T_i, Rl)$ |
| 6          **End If** |
| 7          **If** $T_i$ is an internal node: |
| 8              Mark $T_i$ as *not stale* |
| 9              *ensure_best_splitting*$(T_i \to$ left$)$ |
| 10             *ensure_best_splitting*$(T_i \to$ right$)$ |
| 11         **End If** |
| 12     **End If** |
| 4  **End Function** |

visit every node and to carry out an explicit check. However, it is not practical to visit all the decision nodes for splitting-rule evaluation. A decision node and its subtrees do not necessitate a change if no new cases are added to the node. A *stale* marker is thus defined to denote whether the score on a node is changed due to the addition of a case (see Fig. 4.10). If a node is not *stale*, no change will be incorporated on the subtrees of the node. The use of a *stale* marker helps reduce the computation complexity of incremental training of a decision tree.

Figure 4.9 shows the procedure *ensure_best_splitting*. Consider the tree on the left side of Fig. 4.10 as an example, we carry out the calculation of information gains for each syndrome, 1.68 for $T_1$, 0.45 for $T_3$, and 1.93 for $T_4$, respectively. Therefore, the syndrome $T_4$ with the highest information gain is selected and pulled up to be the new root node. The child nodes of new root node $T_4$ must therefore be modified to adapt to the new rule, and recursive changes in the tree structure are necessary.

## 4.3.4 Tree Transposition

Rebuilding the entire tree can be prohibitively time-consuming and the past cases may no longer be available for rebuilding the DT. It is therefore desirable that the tree be reconfigured at low cost. This change can be accomplished through a sequence of recursive tree transpositions.

Consider the tree transposition on the decision tree as shown in Fig. 4.10; a node of $T_4$ is elevated while a node of $T_1$ descends down the tree. Note that the only change is that the subtrees of two nodes of $T_4$ are reattached to the new nodes of $T_1$, thus the scores on the old nodes $T_1$ and $T_4$ are updated. The score of the left branch of the new node of $T_4$ is the sum of the scores of the left branches of the two previous nodes corresponding to $T_4$. Meanwhile, the score of the right branch of the new node of $T_4$

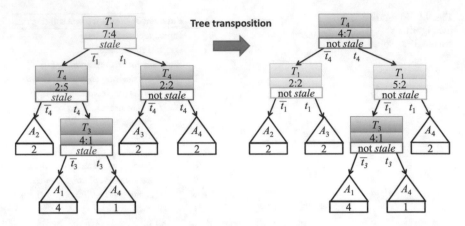

**Fig. 4.10** Illustration of the tree transposition procedure

is the sum of the scores of the right branches of the two previous nodes corresponding to $T_4$. Moreover, there is no need to conduct tree transposition on the node of $T_3$, since $T_3$ is still the best splitting rule on that decision node.

## 4.4 Diagnosis Flow Based on Incremental Decision Trees

The iDT-based diagnosis flow consists of the four steps shown in Fig. 4.11. In general, a set of syndromes and corresponding repair actions are first defined by test designers. Next, an initial decision tree is derived for diagnosis. It can be either written by a test designer or trained from a board-repair database. The DT-based diagnosis system is ready for the diagnosis of new cases. Those new cases in turn can be used to update the diagnosis system.

**Step 1**:    We first define the format of data, including inputs and outputs of DTs. This information is derived from board design and test design. The fault syndromes are defined and recorded in the log files for extraction as the inputs, and the corresponding repair component is taken as the output. Details of the data-preparation stage have been described in Sect. 4.1.

**Step 2**:    In this step, we can either start from a blank tree, which has no diagnosis ability or start with an existing tree. The starting tree can be derived by an experienced test designer, who is familiar with the diagnosis procedure. The scores in each node can be assigned weights according to the confidence level of the test designer. If the test designer has a low level of confidence, he can assign lower weights and let the tree be updated via automated learning during volume manufacturing. In contrast, if the initial weights are set large, the decision tree is more rigid and hard to alter. We compare the different weight settings of a decision tree generated by test designers in Sect. 4.5.

**Fig. 4.11** The diagnosis
flow using DTs

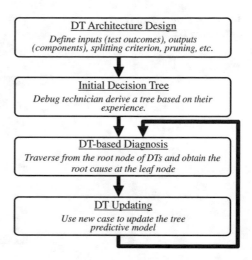

**Step 3**: We can use the DT-based diagnosis system for new-case diagnosis. The new-case diagnosis procedure is shown in Fig. 4.12. We start the diagnosis from the root node of the DT, which is the most discriminative syndrome. Based on the observation of this syndrome, we determine the next syndrome for diagnosis. The follow-up procedure is conducted in a heuristic and recursive manner in that the selection of a new syndrome is determined from the observed result of the current test. Eventually, a leaf node is reached after analysis of several syndromes as part of the DT that is traversed. We can predict the root cause based on the suggestion of the leaf node.

**Step 4**: The existing decision tree can be updated by incorporating the new cases. If a new case alters the decision tree based on the evaluation of the best-splitting rule, a tree transposition is necessary. The decision tree follows the procedure described in Sect. 4.3 to conduct a recursive tree transposition from top to bottom. The training procedure can be implemented using existing machine learning algorithms available in the Weka toolkit [7].

To ensure the accuracy of the proposed DT-based diagnostic system, we consider up to three attempts for identifying root-cause candidates for repair. A board must be scrapped after three unsuccessful attempts at repair. The first attempt corresponds to the first leaf node reached during diagnosis. When the first attempt fails to repair the board, we seek the second root-cause candidate either from the minority class labels in the impurity leaf node or from the sub-tree of the first attempt's parent node. This means that we assign an observation error on the parent node (syndrome) of the first attempt since this parent node led to a wrong repair attempt. By continuing the diagnosis procedure using the DT, we will reach another leaf node corresponding to a root-cause candidate that is different from that for the first attempt. The third attempt follows the steps described above.

**Fig. 4.12** Diagnosis of new cases using DTs

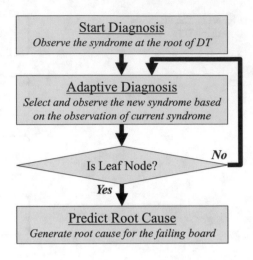

## 4.5 Results

Experiments were performed on two industrial boards that are currently in high-volume production. Relevant information about the boards is provided in Table 2.1, as reproduced below in Table 4.1.

Algorithms were implemented using MATLAB 2011b statistics toolbox [8]. Experiments were run on a 64-bit Linux system with 12 GB of RAM and quad-core Intel i7 processors running at 2.67 GHz. Diagnosis results were obtained for different designs of the DTs, e.g., for various splitting criteria, values of the pruning parameter. Diagnosis results for DT-based system using information gain as the splitting criterion and a relatively small value for the pruning parameter provides higher diagnosis accuracy and leads to fewer syndromes for diagnosis.

In this book, we use the same *cross-validation* methods and concepts of success ratios, as described in Chap. 2, to evaluate the performance of ANN-based, SVM-based, and DT-based diagnosis systems.

**Table 4.1** Information about the industrial boards used for classification and the log data available

|  | Board 1 | Board 2 |
|---|---|---|
| Number of syndromes | 546 | 375 |
| Number of repair candidates (components) | 153 | 116 |
| Number of boards | 1613 | 1351 |

## 4.5.1  Evaluation of DT-Based Diagnosis System

We first compare the number of syndromes $n_T$ used in DTs and average number of syndromes $n_A$ used for identifying a root cause using DTs to those using ANNs and SVMs; see Table 4.2. As described in previous chapters, ANNs and SVMs use all the available syndromes for training and new-case diagnosis; the number of syndromes used for diagnosis is therefore equal to that used for training. In contrast, DTs require even fewer syndromes in diagnosis compared to the reduced number needed for training. For example, after DT training, the number of syndromes for diagnosis is reduced from 546 to 155 or less for Board 1 and from 375 to 83 for Board 2. Different splitting criteria lead to different structures of DTs and therefore different numbers of syndromes. For example, the number of syndromes used for training for Board 1 is 151 using Gini index, 155 using information gain, and 153 using Twoing. The average number of syndromes (average over all the failing boards) for identifying a root cause is only 9 for Board 1 and only 8 for Board 2.

We next compare the SRs for DTs to that for SVMs and ANNs in Table 4.2. We find that the SRs for DTs remains significantly high even when the number of syndromes used for diagnosis are reduced by an order of magnitude for three boards. Our results show that we can identify a small subset of syndromes through DT training that has the same discriminative ability as the complete set of syndromes for diagnosis and

**Table 4.2** Comparison of SRs and number of syndromes obtained by ANNs, SVMs and DTs with different splitting criteria

| Machine learning method | | $SR_1$ (%) | $SR_2$ (%) | $SR_3$ (%) | $n_T$ (%) | $n_A$ (%) |
|---|---|---|---|---|---|---|
| (a) Board 1 | | | | | | |
| Decision trees | Gini index Info. gain Twoing | 65.0 | 75.0 | 81.5 | 151 | 9 |
| | | 65.0 | 75.0 | 82.5 | 155 | 9 |
| | | 62.5 | 72.0 | 79.0 | 153 | 9 |
| Baseline methods | ANNs | 67.9 | 78.1 | 84.4 | 546 | 546 |
| | SVMs | 73.2 | 80.4 | 88.2 | 546 | 546 |
| (b) Board 2 | | | | | | |
| Decision trees | Gini index Info. gain Twoing | 52.7 | 68.0 | 72.7 | 83 | 8 |
| | | 55.3 | 69.0 | 72.7 | 83 | 8 |
| | | 52.7 | 68.0 | 73.7 | 83 | 8 |
| Baseline methods | ANNs | 57.9 | 70.3 | 75.1 | 375 | 375 |
| | SVMs | 66.3 | 74.3 | 84.1 | 375 | 375 |

by ANNs or SVMs. We also find that the SRs are higher using Information Gain as the splitting criterion.

We next compare the SRs achieved using DTs with that achieved using the ANN and SVM baseline methods, when the set of syndromes is the same for all these methods. The results are shown in Tables 4.3 and 4.4. The sets of syndromes selected by DTs are used for training SVM and ANN models. We also train the SVMs (ANNs) with randomly selected syndromes, such that the number of syndromes in the SVMs (ANNs) is same as in the DTs. Random selections are repeated 20 times. The maximum, minimum, and average SRs obtained are listed in Tables 4.3 and 4.4. These results show that for the same small number of syndromes, DT outperforms ANNs and SVMs in nearly all cases. Moreover, DTs can be used to select the most discriminative syndromes for use by ANNs and SVMs—as a result, the success ratio is often higher than when a much larger number of syndromes is used.

**Table 4.3** Comparison of DTs and SVMs with reduced set of syndromes

|          | DTs (%) | SVMs[a] (%) | SVMs[b](min) (%) | SVMs[b](max) (%) | SVMs[b](avg.) (%) |
|----------|---------|-------------|------------------|------------------|-------------------|
| (a) Board 1 |      |             |                  |                  |                   |
| $SR_1$   | 65.0    | 73.2        | 35.5             | 54.7             | 43.9              |
| $SR_2$   | 75.0    | 80.4        | 51.7             | 68.2             | 60.0              |
| $SR_3$   | 82.5    | 88.2        | 65.8             | 78.3             | 72.8              |
| (b) Board 2 |      |             |                  |                  |                   |
| $SR_1$   | 55.3    | 66.3        | 29.2             | 37.5             | 31.1              |
| $SR_2$   | 69.0    | 74.3        | 43.3             | 61.7             | 56.8              |
| $SR_3$   | 72.7    | 84.1        | 57.5             | 70.0             | 63.3              |

[a] SVMs trained using syndromes from DTs
[b] SVMs trained using randomly selected $n$ syndromes, where $n$ is the number of syndromes in the DT

**Table 4.4** Comparison of DTs and ANNs with reduced set of syndromes

|          | DTs (%) | ANNs[c] (%) | ANNs[d](min) (%) | ANNs[d](max) (%) | ANNs[d](avg.) (%) |
|----------|---------|-------------|------------------|------------------|-------------------|
| (a) Board 1 |      |             |                  |                  |                   |
| $SR_1$   | 65.0    | 67.9        | 16.3             | 23.8             | 19.0              |
| $SR_2$   | 75.0    | 78.1        | 19.2             | 31.3             | 24.5              |
| $SR_3$   | 82.5    | 84.4        | 25.5             | 41.7             | 29.1              |
| (b) Board 2 |      |             |                  |                  |                   |
| $SR_1$   | 55.3    | 57.9        | 14.2             | 25.0             | 20.4              |
| $SR_2$   | 69.0    | 70.3        | 19.7             | 33.3             | 24.2              |
| $SR_3$   | 72.7    | 75.1        | 29.2             | 37.5             | 33.7              |

[c] ANNs trained using syndromes from DTs
[d] ANNs trained using randomly selected $n$ syndromes, where $n$ is the number of syndromes in the DT

## 4.5.2 Evaluation of Incremental DT-Based Diagnosis System

Next, we evaluate the benefit of using incremental learning in diagnosis systems. In Figs. 4.13–4.14, we compare the diagnosis accuracies of four diagnosis systems. *DT* is a diagnosis system based on the DT written by the debug technicians. The diagnosis system is nonincremental. *iDT* is a diagnosis system based on an incremental DT, but the DT starts with an empty tree at the beginning of volume manufacturing. $iDT^+$ and $iDT^-$ are two iDT-based diagnosis systems with initial DT derived from an experienced technician. The difference is that the scores of all leaf nodes in $iDT^+$ are initially assigned high (e.g., 50), while those of $iDT^-$ are assigned low scores (e.g., 1). We observe that the diagnosis accuracies of the *DTs*, i.e., $SR_1$ are approximately 40–50 % on average, which is common in modern product lines. As the product volume increases, the diagnosis accuracy does not increase, but even drops sometimes. *iDT* shows very low diagnosis ability at the beginning of volume production, but the diagnosis accuracy ramps up gradually due to incremental learning as the product volume accumulates. Next, we observe that the diagnosis accuracies of $iDT^-$ and $iDT^+$ are significantly better. At the beginning, the performances of $iDT^-$ and $iDT^+$ are close to *DT*, and then the diagnosis accuracies increase as more boards are available for learning. Note that $iDT^-$ performs much better than $iDT^+$ in all the three diagnosis systems. The reason is that $iDT^-$ is easier to adapt to the knowledge of new error scenarios.

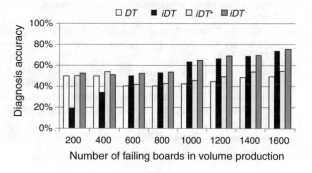

**Fig. 4.13** Comparison of fault diagnosis using batch DTs, incremental DTs, incremental DTs* for Board 1

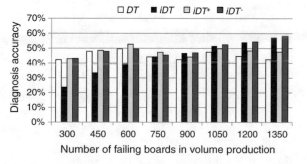

**Fig. 4.14** Comparison of fault diagnosis using batch DTs, incremental DTs, incremental DTs* for Board 2

## 4.6   Chapter Summary

This chapter presented an adaptive board-level functional fault diagnosis method based on incremental decision trees (iDTs). It is shown that how faulty components can be classified based on the discriminative ability of the syndromes in DT training. The diagnosis procedure has been constructed as a binary tree, with the most discriminative syndrome as the root and the final repair suggestions are available as the leaf nodes of the tree. In this way, accurate diagnosis can be carried out at low cost with a small number of syndromes. Furthermore, we have combined the diagnosis knowledge from debug technicians with the knowledge gained from failing boards during volume manufacturing. Therefore, the problem of low diagnosis accuracy at the beginning of volume production can be alleviated. The diagnosis system can adapt to occurrences of new error scenarios to enhance diagnosis accuracy on-the-fly. Diagnosis results for two complex boards from industry, currently in volume production, highlight the effectiveness of the proposed approach.

# References

1. Breiman L (1984) Classification and regression trees. Chapman & Hall/CRC, Boca Raton
2. Quinlan J (1986) Induction of decision trees. Mach Learn 1(1):81–106
3. Zhang Z, Chakrabarty K, Wang Z, Wang Z, Gu X (2011) Smart diagnosis: efficient board-level diagnosis and repair using artificial neural networks. In: Proceedings IEEE International Test Conference (ITC), pp 1–10
4. Zhang Z, Gu X, Xie Y, Wang Z, Chakrabarty K (2012) Diagnostic system based on support-vector machines for board-level functional diagnosis. In: Proceedings IEEE European Test Symposium (ETS), pp 1–6
5. Vlahou A, Schorge JO, Gregory BW, Coleman RL (2003) Diagnosis of ovarian cancer using decision tree classification of mass spectral data. J Biomed Biotechnol 2003(5):308–314
6. Abu-Hakima S (1992) Visualizing and understanding diagnoses. Can Artif Intell 30:1–10
7. McLachlan G, Do K, Ambroise C (2004) Analyzing microarray gene expression data, vol 422. Wiley, New York
8. Matlab Statistics Toolbox (2013) http://www.mathworks.com/products/statistics/

# Chapter 5
# Information-Theoretic Syndrome and Root-Cause Evaluation

Reasoning-based methods are promising because a detailed system model is not needed to construct the diagnosis system [1, 2]. The diagnosis engine is incrementally built based on the database of successfully repaired faulty boards. Machine learning techniques enable reasoning-based diagnosis, leveraging ease of implementation, high diagnosis accuracy, and continuous learning. Repairs of faulty components are suggested through a ranked list of suspect components, e.g., based on artificial neural networks (ANNs) and support-vector machines (SVMs) [3]. However, a machine learning-based diagnosis system requires a sizeable database for training. Ambiguous root-cause identification may result if the diagnosis system lacks a sufficient database with adequate information for mapping errors to root causes. There is a need for a rigorous framework that can evaluate an existing diagnosis system using quantifiable metrics, and provide means for enhancing the accuracy of diagnosis. Ambiguity in root-cause identification confounds diagnosis, hence such ambiguities need to be identified and used as a guideline for test development. The accuracy of diagnosis and time needed for diagnosis also depend on the quality of syndromes (erroneous observations). Redundant or irrelevant syndromes not only lead to long diagnosis time, but also increase diagnosis complexity.

In this chapter, we propose an evaluation and enhancement framework for a functional fault diagnosis system, as shown in Fig. 5.1. This framework is based on information-theoretic analysis. The discriminative ability of each root cause is calculated, which allows us to screen out root causes with low discriminative ability, and identify ambiguous root-cause pairs. The evaluation system also carries out rigorous syndrome analysis. A subset of representative syndromes is selected with maximum relevance and minimum redundancy. The irrelevant or redundant syndromes can subsequently be analyzed to derive better tests for diagnosis.

The remainder of this chapter is organized as follows. Section 5.1 presents the problem and describes the work contributions. Section 5.2 describes the evaluation framework based on subset-selection and class-relevance statistics. Section 5.3 describes the diagnosis framework that includes the proposed evaluation system.

© Springer International Publishing Switzerland 2017                                                      79
F. Ye et al., *Knowledge-Driven Board-Level Functional Fault Diagnosis*,
DOI 10.1007/978-3-319-40210-9_5

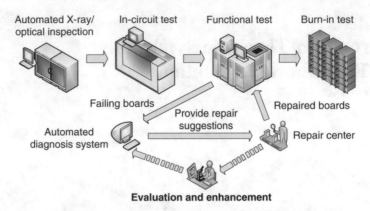

**Fig. 5.1** Overview of the diagnosis system with self-evaluation

Section 5.4 presents results for two industrial boards at volume production. Compared to traditional metrics such as overall success ratios, the proposed evaluation framework provides detailed analysis for each syndrome and each root cause with respect to their contribution to prediction accuracy. Finally, Sect. 5.5 concludes the chapter.

## 5.1  Background and Chapter Highlights

Machine learning provides an unprecedent opportunity for developing an automated functional fault diagnosis system that can become more effective over time through self-learning. The use of an automated diagnosis system involves less human effort and can provide accurate diagnostic guidance. The diagnosis system can be built through learning from historical repair data, which determines the prediction performance for future faulty boards. However, the quality of historical data varies across different types of boards and different stages of diagnosis system deployment. In prior work, evaluation of a diagnosis system has been based on success ratio (SR), which is defined as the ratio of the number of correctly diagnosed boards to the total number of failed boards. However, SR only offers an overall performance metric for the diagnosis system without providing any guidance for diagnosis system improvement. It is therefore necessary to develop an evaluation framework for the diagnosis system, which can provide valuable feedback for selecting syndromes, designing tests, and reasoning about root-cause ambiguities.

We propose a new evaluation framework, as shown in Fig. 5.2, that leverages two analysis methods, targeting syndromes and root-cause components, respectively. Syndrome analysis based on subset selection is used to select an effective, but reduced, set of syndromes for use by a diagnosis system, such as based on SVMs [3]. Root-cause analysis relies on information-theoretic metrics such as *precision*

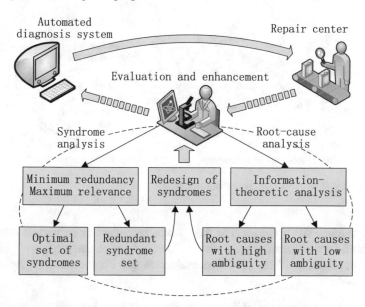

**Fig. 5.2** Evaluation framework for a diagnosis system and the targeted problems

and *recall*. These metrics measure the discriminative ability of a diagnosis system to differentiate a root cause from other candidates. The proposed frameworks aims to provide guidelines for addressing the following practical problems:

**Lack of experience for particular root causes**: It is often the case that a diagnosis system is initially deployed with a limited number of faulty boards and insufficient history data. Boards with a particular root cause may rarely occur in the corresponding history cases. During subsequent diagnosis, if we encounter a board with such a root cause, the diagnosis system will not provide any useful information. The identification of these root causes in an early stage of diagnosis system deployment will facilitate the evaluation of the system and point out weaknesses.

**Ambiguous root causes**: The components at board level tend to be coupled to each other, e.g., through layout proximity and signal flow. Given a set of syndromes, a diagnosis system may suggest several potential root causes with the same level of confidence. One reason for such ambiguity is that functional-test packages are typically designed by multiple teams. It can be hard to isolate root causes from each other due to the lack of some key syndromes. It is therefore desirable that the evaluation framework reveal these coupled components and provide guidelines for generating new tests.

**Redundant or irrelevant syndromes**: Although multisite design collaboration can reduce product release time and boost productivity, syndrome sets from different teams may overlap. Redundant syndromes not only add to diagnosis cost, but also increase diagnosis complexity. A representative set of syndromes can be extracted

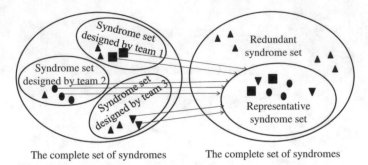

**Fig. 5.3**  Selection of a representative syndrome subset

from the original syndrome set (Fig. 5.3), while the remaining redundant syndromes can be used for careful analysis at various levels of design abstraction.

The contributions of the proposed evaluation and enhancement framework are listed below:

- A method to analyze the ability of the diagnosis system to discriminate among root causes.
- A technique to analyze and optimize a syndrome set to obtain a small subset with high discriminative ability.
- Guidelines for test-design teams so that information about redundant syndromes and ambiguous pairs of root causes can be used for future test optimization.

## 5.2  Evaluation Methods for Diagnosis Systems

In this section, we describe the proposed techniques for evaluating a reasoning-based diagnosis system.

### 5.2.1  Subset Selection for Syndromes Analysis

Subset selection (also known as feature selection) [4, 5] aims to identify a set of most important features for characterization. Suppose we have a database $C$ of successfully repaired faulty boards, which have $N$ potential root-cause candidates denoted by the elements of the set $\mathbf{A} = \{A_1, A_2, \ldots, A_N\}$. To diagnose the root cause for a faulty board, we start with $M$ syndromes, i.e., $\mathbf{S} = \{S_1, S_2, \ldots, S_M\}$. The feature selection problem is to find $m$ syndromes, $m \leq M$, that optimally characterize $\mathbf{A}$ (see right part of Fig. 5.3). The optimal characterization condition involves the determination of the minimum classification error.

One of the most popular approaches to the subset-selection problem is based on the rule of max-relevance (MaxRel) feature selection. The goal is to select syndromes with the highest relevance to the root cause $A_i$. Relevance is characterized in terms of mutual information, which is one of the widely used metrics to define dependency. For each target root cause $A_i$ and a given syndrome $S_j$, their mutual information $E(A_i, S_j)$ is calculated as:

$$E(A_i, S_j) = -p(A_i|s_j) \log p(A_i|s_j) - p(A_i|\bar{s}_j) \log p(A_i|\bar{s}_j), \qquad (5.1)$$

where the logarithmic base is 2, $s_j$ is the event that syndrome $S_j$ manifests itself, and $\bar{s}_j$ is the complementary event. The measure $p(A_i|s_j)$ is the probability of $A_i$ being the root cause if $S_i$ manifests itself. We then calculate the relevance value $D(\mathbf{A}, S_j)$ between the syndrome $S_j$ and root-cause set $\mathbf{A}$ as follows:

$$D(\mathbf{A}, S_j) = \frac{1}{|\mathbf{A}|} \sum_{A_i \in \mathbf{A}} E(A_i, S_j). \qquad (5.2)$$

The syndromes in $\mathbf{S}$ can be ranked according to $D(\mathbf{A}, S_j)$ values in descending order, where $D(\mathbf{A}, S_1') \geq D(\mathbf{A}, S_2') \geq \cdots \geq D(\mathbf{A}, S_M')$. The MaxRel set $\mathbf{S}' = \{S_1', S_2', \ldots, S_m'\}$ is a selected subset of the top $m$ syndromes from $\mathbf{S}$ according to the ranked syndrome list, where $1 \leq m \leq M$ and $\mathbf{S}' \subseteq \mathbf{S}$ [4]. We can determine the best choice of $m$ by using a heuristic to evaluate the diagnosis system with different syndrome sets $\mathbf{S}'$ (each with a different value of $m$) in terms of diagnosis accuracy. However, in feature selection, it is recognized that the combinations of individually good features do not necessarily lead to effective classification. In other words, "the $m$ best syndromes are not the best $m$ syndromes" [6, 7]. To reduce the redundant syndromes in the syndrome set $\mathbf{S}'$, $\mathbf{S}'$ is therefore further evaluated by computing its redundancy value $R(\mathbf{S}')$ as shown below:

$$R(\mathbf{S}') = \frac{1}{|\mathbf{S}'|^2} \sum_{S_i' \in \mathbf{S}'} \sum_{S_j' \in \mathbf{S}'} E(S_i', S_j'), \qquad (5.3)$$

where $E(S_i', S_j')$, $S_i' \neq S_j'$, is the mutual information between $S_i'$ and $S_j'$, which is similar to the definition of $E(A_i, S_j)$, i.e., $E(S_i', S_j') = -p(s_i|s_j) \log p(s_i|s_j) - p(s_i|\bar{s}_j) \log p(s_i|\bar{s}_j) - p(\bar{s}_i|s_j) \log p(\bar{s}_i|s_j) - p(\bar{s}_i|\bar{s}_j) \log p(\bar{s}_i|\bar{s}_j)$. We then calculate the minimum-redundancy maximum-relevance (mRMR) value for each $\mathbf{S}'$, as follows:

$$mRMR(\mathbf{S}') = \frac{1}{|\mathbf{S}'|} \sum_{S_k' \in \mathbf{S}'} D(\mathbf{A}, S_k') - R(\mathbf{S}'). \qquad (5.4)$$

We can next determine the minimum-redundancy-maximum-relevance (mRMR) syndrome subset $\mathbf{S}^*$ with the largest mRMR value, as follows:

$$\mathbf{S}^* = \max_{\mathbf{S}'} \{mRMR(\mathbf{S}')\}. \qquad (5.5)$$

We can now evaluate the diagnosis system with the mRMR syndrome set $\mathbf{S}^*$ in terms of diagnosis accuracy. The subset of syndromes thus selected leads to minimum classification error.

## 5.2.2   Class-Relevance Statistics

Success ratio (SR) has been used for evaluating board-level diagnosis systems in prior work [3]. However, SR is a coarse metric for multiclass classifiers (e.g., a board-level functional fault diagnosis system) and does not provide any suggestion for improving diagnosis. We introduce a fine-grained set of information-theoretic metrics, which are used for comprehensively evaluating the diagnosis system. First, we use the concept of a *confusion matrix*, a table layout that allows visualization of the performance of a classifier. Each column of the matrix represents the cases in a predicted root cause, and each row represents the cases of an actual root cause. The name stems from the fact that it makes it easy to see if the system is confusing between root causes. For example in Table 5.1, we form a representative binary classification system with two classes encoded as positive and negative, respectively. The entries in the confusion matrix have the following meaning:

- TP is the number of correctly predicted positive cases;
- FP is the number of incorrectly predicted positive cases;
- FN is the number of incorrectly predicted negative cases;
- TN is the number of correctly predicted negative cases.

Two popular metrics (*precision* and *recall*) make up the proposed root-cause analysis framework.

Positive predictive value (PPV), also known as *precision*, is the proportion of the predicted positive cases that are correct, calculated using the equation [7]:

$$PPV(precision) = \frac{TP}{TP + FP} \tag{5.6}$$

True positive rate (TPR), also known as *recall*, is the proportion of positive cases that are correctly identified, calculated as follows:

$$TPR(recall) = \frac{TP}{TP + FN} \tag{5.7}$$

**Table 5.1**  Confusion matrix

|              |          | Predicted class | |
|--------------|----------|----------|----------|
|              |          | Positive | Negative |
| Actual class | Positive | TP       | FN       |
|              | Negative | FP       | TN       |

Note that the SR metric can be expressed in terms of *recall*. It is the sum of all the correctly predicted cases for each root cause $A_i \in \mathbf{A}$, calculated as SR = $\sum_{A_i \in \mathbf{A}} \text{TPR}_{A_i} \cdot n_{A_i}$.

A *precision* score of 1.0 for a root cause $A_i$ means that all predictions about positive cases are correct, i.e., there are no false positives. On the other hand, a *recall* score of 1.0 means that all the cases with root cause $A_i$ are correctly predicted, i.e., there are no false negatives. For example, in board-level diagnosis, *precision* describes the percentage of success in predicting a root cause, while *recall* reflects the percentage of success for a root cause to be predicted. One metric alone provides only partial information, limited to either prediction or reality. A combination of these two metrics provides a more complete picture. Hence, compared to the traditional metric (SR), we now have a more comprehensive set of metrics, through which we can obtain more information on prediction accuracy. In addition, we also log all ambiguous pairs of root causes, which can be referred to design teams for detailed analysis at various levels of design.

## 5.3 Evaluation and Enhancement Framework

Prediction accuracy improves when the volume of successfully repaired faulty boards increases [3]. However, in practice, the database is limited in size and the diagnosis system suffers from low prediction accuracy due to the lack of availability learning data. The diagnosis system must therefore be assessed periodically during the deployment to measure its prediction quality.

### 5.3.1 Evaluation and Enhancement Procedure

Given a database of failed boards, syndrome and root-cause analyses are processed in parallel in the evaluation framework (see Fig. 5.2). The syndrome analysis is based on the minimum redundancy and maximum relevance (mRMR) method. Mutual information between each syndrome and root cause is iteratively calculated to form a set of syndromes with maximum relevance, followed by syndrome reduction from this set based on the calculation of relevance between syndromes. The set of redundant syndromes can then be provided to the syndrome design team for further analysis based on similarity among functional tests. If some of the redundant syndromes are still considered to be useful by the design team, we regard them as "suspicious" syndromes. The redundant syndromes are removed from the new syndrome set for diagnosis, but the "suspicious" ones will be retained in the diagnosis system for future evaluation.

As discussed in Sect. 5.2, root-cause analysis is based on information-theoretic metrics. First, an $N \times N$ confusion matrix is generated, where $N$ is the number of root causes. For each root cause $A_i$, we define a hypothetical root cause

$A_i^* = \mathbf{A} - \{A_i\}$. We then derive a two-class confusion matrix of $A_i$ and $A_i^*$ from the original confusion matrix. If the *precision* of $A_i$ is less than one, for each faulty board, we create ambiguous root cause pair $\langle A_i, A_j \rangle$, where $A_j$ is the root cause for a board incorrectly predicted to be $A_i$. On the other hand, if the *recall* of $A_i$ is less than one, for each faulty board, we form the root cause pair $\langle A_k, A_i \rangle$, where $A_k$ is the incorrect root cause prediction for a board with the root cause $A_i$. These ambiguous pairs and associated faulty boards will then be provided to the syndrome design team. New tests (resulting in new syndromes) can be developed to split these ambiguous pairs and they can be added into the new syndrome set to improve the diagnosis system. The outputs of this evaluation and enhancement framework include:

- An optimal set of syndromes that contribute to root cause isolation;
- A set of redundant or irrelevant syndromes to be discarded based on the outcome of functional analysis;
- A group of root causes that can be isolated with high confidence;
- A set of ambiguous root-cause pairs, and associated boards for detailed functional analysis.

Based on the outputs of the proposed evaluation framework, new syndromes can be designed to increase the discriminative ability for ambiguous root-cause pairs, and redundant or irrelevant syndromes can be discarded to reduce diagnosis complexity. Based on the new syndrome set, a new diagnosis system can be trained using methods such as fault-insertion tests, FPGA-based board test, model-based reasoning, etc. [8, 9].

### 5.3.2 An Example of the Proposed Framework

Consider a hypothetical demonstration board with five cases that are used for evaluation. Let $x_1, x_2, x_3$, and $x_4$ be four syndromes. If the syndrome manifests itself, we record it as 1, and 0 otherwise. Let us suppose that the board has two candidate root causes A and B, and we encode them as $y = P$ and $y = N$, respectively. In a real scenario, fault syndromes vary across products and tests. Here, we merge the syndromes and the known root causes in one matrix $\mathscr{A} = (\mathscr{B}|\mathscr{C})$, where the left ($\mathscr{B}$) side refers to the syndromes, while the right side ($\mathscr{C}$) refers to the corresponding fault classes. This matrix represents the database information for evaluation.

$$\mathscr{A} = [\mathscr{B}|\mathscr{C}] = \begin{bmatrix} 1 & 1 & 0 & 0 & \vdots & P \\ 1 & 1 & 0 & 1 & \vdots & P \\ 1 & 1 & 0 & 0 & \vdots & P \\ 0 & 1 & 1 & 1 & \vdots & N \\ 0 & 0 & 1 & 1 & \vdots & N \end{bmatrix} \tag{5.8}$$

First, we consider syndrome analysis based on the database. The two-step mRMR method consists of calculating relevance scores and mRMR scores. The mutual information for root cause $P$ are calculated to be 0 for $x_1$, 0 for $x_3$, $-0.27$ for $x_2$, and $-0.38$ for $x_4$ in a descending order, using Eq. (5.2). We can select a subset of one syndrome $\{x_1\}$ with the relevance score of 0. If we select a subset of two syndromes, we can select $\{x_1, x_2\}$ with the relevance score of $(0 - 0.27)/2 = -0.13$, or $\mathbf{S}' = \{x_1, x_3\}$ with the relevance score of $(0+0)/2 = 0$ and so on. However, we observe that $x_3$ is the complement of $x_1$. To minimize redundancy in the subset, we then calculate the redundancy score to get the mRMR score using Eq. (5.4). The redundancy score $R(\mathbf{S}') = \frac{1}{4}E(x_1, x_3) = 0$, thereby resulting in the $mRMR(\mathbf{S}') = 0 - 0 = 0$ for $\mathbf{S}'$. Therefore, $x_3$ is considered to be redundant if $\{x_1\}$ can be used for classification. Consider two other two-class subsets of syndromes, namely $\{x_1, x_2\}$ with $mRMR = -0.07$, and $\{x_1, x_4\}$ with $mRMR = -0.10$. Based on these numbers, we still choose a one-syndrome subset $\{x_1\}$ with higher $mRMR$ score of 0 for diagnosis. To ensure that $\{x_1\}$ is the minimal syndrome subset, we also use SR to estimate the performance of the diagnosis system using this syndrome subset. We observe that the SRs obtained from a diagnosis system using other multisyndrome sets are no better than using $\{x_1\}$. We then select the optimal syndrome set $\mathbf{S}^* = \{x_1\}$ for this diagnosis system.

Next, we consider root-cause analysis. Suppose the prediction of a diagnosis system for these five boards is $\{P, P, N, N, N\}$, wherein the third board with root cause $P$ is predicted to be $N$, which is incorrect. In the confusion matrix, TP = 2, FP = 1, FN = 0, and TN = 2. Thus, *precision* of $P$ is calculated to be 0.66, which means two out of three predictions of root cause $P$ are correct. In addition, *recall* of $P$ is 1, which mean all the boards with root cause $P$ are correctly predicted.

## 5.4 Results

We conducted the experiments on three industrial boards that are currently in high-volume production. Here, we describe the experimental setup and then present the improvements achieved by using the evaluation framework.

Both syndrome and root-cause analysis are implemented in C/C++. Syndrome analysis is implemented using minimum-redundancy-maximum-relevance toolkits [4]. Note that the proposed evaluation framework is of general use to any diagnosis system. We illustrate the use of the evaluation framework using two examples, i.e., a support-vector machine (SVM)-based diagnosis system and an artificial neural network (ANN)-based diagnosis system, respectively [3].

We use two industrial boards from Table 2.1 to illustrate the evaluation framework. Information about these two industrial boards, diagnosis results before evaluation, and diagnosis results after evaluation and test redesign are reproduced and shown in Table 5.2 for an SVM-based diagnosis system and Table 5.3 for an ANN-based diagnosis system. For example, a total of 1613 repaired boards are collected before evaluation from the contract manufacturer's database for Board 1. A total of 546 fault syndromes are extracted from the failure logs. The number of faulty components identified in the database for repair action is 153. After we apply the evaluation framework on Board 1, additional 256 boards are evaluated.

**Table 5.2** Diagnosis results before and after evaluation for a diagnosis system based on SVMs

|                         | Board 1 | Board 2 |
|-------------------------|---------|---------|
| Pre-evaluation          |         |         |
| Number of syndromes     | 546     | 375     |
| Number of components    | 153     | 116     |
| Number of boards        | 1613    | 1351    |
| $SR_1$                  | 73.2%   | 66.3%   |
| $SR_2$                  | 80.4%   | 74.3%   |
| $SR_3$                  | 88.2%   | 84.1%   |
| Post-evaluation         |         |         |
| Number of syndromes     | 299     | 201     |
| Number of components    | 153     | 116     |
| Number of boards        | 256     | 224     |
| $SR_1$                  | 86.1%   | 89.6%   |
| $SR_2$                  | 90.8%   | 90.9%   |
| $SR_3$                  | 93.2%   | 92.0%   |

**Table 5.3** Diagnosis results before and after evaluation for a diagnosis system based on ANNs

|                         | Board 1 | Board 2 |
|-------------------------|---------|---------|
| Pre-evaluation          |         |         |
| Number of syndromes     | 546     | 375     |
| Number of components    | 153     | 116     |
| Number of boards        | 1613    | 1351    |
| $SR_1$                  | 67.9%   | 57.9%   |
| $SR_2$                  | 78.1%   | 70.3%   |
| $SR_3$                  | 85.4%   | 75.1%   |
| Post-evaluation         |         |         |
| Number of syndromes     | 275     | 202     |
| Number of components    | 153     | 116     |
| Number of boards        | 256     | 224     |
| $SR_1$                  | 81.2%   | 76.4%   |
| $SR_2$                  | 86.5%   | 87.5%   |
| $SR_3$                  | 93.2%   | 89.3%   |

To ensure real-time diagnosis and repair, we assume that we are allowed at most three attempts to replace the potential failing components. We define $SR_1$ as the success ratio corresponding to the case that the board is deemed to have been successfully repaired only when the actual faulty component is identified and placed at the top of the list of root-cause candidates. We also define $SR_2$ ($SR_3$) as the success ratio corresponding to the case that a board is deemed to have been successfully repaired if the actual faulty component is in the first two (three) positions in the list of root-cause candidates.

### 5.4.1 Demonstration of Syndrome Analysis

We first do the syndrome analysis, followed by root-cause analysis. To evaluate the effectiveness and efficiency of syndrome analysis, we examine the minimum number of syndromes used by the diagnosis system, while achieving the same diagnosis accuracy obtained with the entire set of syndromes. Tables 5.2 and 5.3 shows the number of syndromes before and after the syndrome analysis for the three boards for an SVM-based and an ANN-based diagnosis systems, respectively. For example, for Board 1 using SVM, the number of syndromes after analysis is reduced to 299, which is only 54 % of the total number of syndromes before syndrome analysis.

There are several sources of redundancy that contribute to the significant reduction in the number of syndromes:

1. A few syndromes are effective indicators of failing boards but they are insensitive to the specific root cause that results in the board failure. These syndromes are useful for failure screening, but of no help for fault isolation.
2. A number of syndromes have never manifest themselves in any failing scenario thus far. We regard them as "suspicious" syndromes, but there is a large chance that those syndromes will never be manifested in the future.
3. Most syndromes are fine-grained at the pin level of the component. However, those coarse-grained syndromes are adequate for fault isolation at component level. We can combine several syndromes for the same component into one syndrome.

### 5.4.2 Demonstration of Root-Cause Analysis

To evaluate the effectiveness of root-cause analysis, we present the *precision* and *recall* results for the three boards in Figs. 5.4 and 5.5. Most root causes in all boards can be clearly differentiated from others. For example in Board 1 using the SVM-based diagnosis system in Fig. 5.4a, the number of root causes with *precision* of 1.0 is 105, while the number of root causes with *recall* of 1.0 is 83. In contrast, some root causes cannot be differentiated at all, since either their *precision* or *recall* is less than 0.1. There are a total of 22 root causes of low discriminative ability, of which 13 root causes have both low *precision* and low *recall*, 8 root causes only have low *precision*, and rest 7 root causes only have low *recall*. The values of *precision* and *recall* for Board 2 can be seen in Fig. 5.4b.

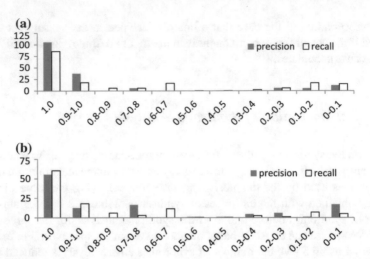

**Fig. 5.4** *Precision* and *recall* distribution for an SVM-based diagnosis system for three boards. **a** Board 1. **b** Board 2

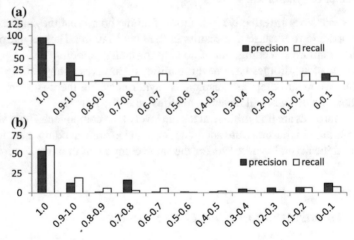

**Fig. 5.5** *Precision* and *recall* distribution for an ANN-based diagnosis system for three boards. **a** Board 1. **b** Board 2

Next, we present additional information about ambiguous root-cause pairs from the faulty boards in Figs. 5.6 and 5.7. For example for Board 1 and using SVM, a total of 60,031 root-cause pairs are formed for 347 root causes. Out of these, 98.5 % of the pairs can be differentiated, while only 1.5 % of the pairs are ambiguous. We found 0.6 % of the pairs are ambiguous between two root causes, 0.1 % of the pairs in a group of three, 0.8 % of the pairs in a group of four, and 0.1 % of the pairs in a group of eight.

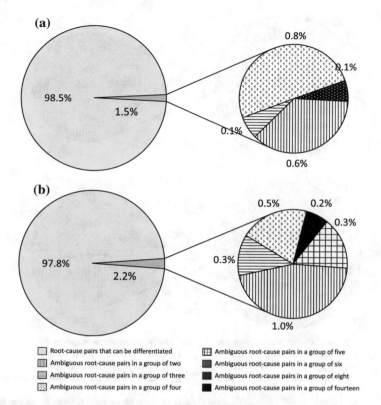

**Fig. 5.6** Comparison of ambiguous root-cause pairs for an SVM-based diagnosis system for three boards. **a** Board 1. **b** Board 2

There are several reasons for these ambiguous root-cause pairs:

1. Memories are paired and occupy one address/data path connected to the chip. Limited observability is a consequence of such a design. In this scenario, external memory BIST can help differentiate between those memories.
2. A number of passive components, oscillators, and frequency dividers form a clock chain/tree, where only limited design-for-tests (DfT) features can be inserted. Such DfT limitations can also be observed in the power-supply tree. In such scenarios, increasing the number of observation points on the board is desirable.
3. At the system boot-up stage, the same syndrome is observed due to any candidate defect from a pool of a number of chips and peripheral clock/voltage components, and passive components. In such a scenario, collecting additional syndromes at the boot-up stage helps increase discriminative ability between root causes.

Last, we demonstrate the effectiveness of the evaluation framework by evaluating the SRs obtained from the updated set of tests, which are guided by the evaluation framework. As shown in Table 5.2 for SVM-based diagnosis systems and Table 5.3 for ANN-based diagnosis systems, all the success ratios increase. For example for

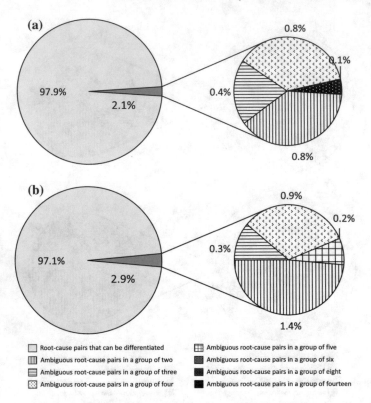

**Fig. 5.7** Comparison of ambiguous root-cause pairs for an ANN-based diagnosis system for three boards. **a** Board 1. **b** Board 2

SVM-based diagnosis system for Board 1, SR1, SR2, and SR3 are 73.2%, 80.4%, and 88.2%, respectively, on a total of 1613 failing boards before the evaluation framework is applied on the diagnosis system. The success ratios increase to 86.1, 92.8, and 93.2% on the new 256 failing boards after we increase the discriminative ability of the diagnosis system by adding new tests. The improvement of diagnosis accuracy can also been seen for ANN-based diagnosis systems.

## 5.5  Chapter Summary

This chapter has proposed an evaluation and enhancement framework based on information-theoretic metrics. Reasoning-based board-level fault diagnosis system requires a rich set of syndromes and a sizable database of faulty boards. However, an insufficient database of faulty boards and redundant syndromes may lead to ambiguous diagnosis, thereby resulting in an increase in diagnosis cost. Syndrome analysis based on subset selection can generate an optimal set of syndromes, and root-cause analysis based on class-relevance metrics can help identify root causes of low discriminative ability. Identifying these redundant syndromes and ambiguous root-cause pairs can provide guidelines for optimizing the syndrome set for diagnosis. Promising experimental results have been obtained from two industry boards in high-volume production.

# References

1. O'Farrill C, Moakil-Chbany M, Eklow B (2005) Optimized reasoning-based diagnosis for non-random, board-level, production defects. In: Proceedings IEEE international test conference (ITC), pp 173–179
2. Fenton W, McGinnity T, Maguire L (2001) Fault diagnosis of electronic systems using intelligent techniques: a review. IEEE Trans Syst Man Cybern Part C: Appl Rev 31:269–281
3. Ye F, Zhang Z, Chakrabarty K, Gu X (2013) Board-level functional fault diagnosis using artificial neural networks, support-vector machines, and weighted-majority voting. IEEE Trans Comput-Aided Des Integr Circuits Syst (TCAD) 32(5):723–736
4. Peng H, Long F, Ding C (2005) Feature selection based on mutual information criteria of max-dependency, max-relevance, and min-redundancy. IEEE Trans Pattern Anal Mach Intell 27(8):1226–1238
5. Webb A (2003) Statistical pattern recognition. Wiley, New Jersey
6. Oja E, Hyvarinen A, Karhunen J (2001) Independent component analysis. Wiley, New York
7. Cover T, Thomas J (2006) Elements of information theory. Wiley-Interscience, New York
8. Zhang Z, Wang Z, Gu X, Chakrabarty K (2012) Physical-defect modeling and optimization for fault-insertion test. IEEE Trans VLSI Syst (TVLSI) 20(4):723–736
9. Manley D, Eklow B (2002) A model based automated debug process. In: IEEE Board Test Workshop, pp 1–7

# Chapter 6
# Handling Missing Syndromes

The diagnosis accuracy of reasoning-based diagnosis engine may be significantly reduced when the repair logs are fragmented and some errors, or syndromes, are not available during diagnosis. Since root-cause isolation for a failing board relies on reasoning based on syndromes, any information loss (e.g., missing syndromes) during the extraction of a diagnosis log may lead to ambiguous repair suggestions.

In this chapter, we propose a board-level diagnosis system with the feature of handling missing syndromes using the method of imputation. The syndromes from a faulty-board log are analyzed and *imputed* with appropriate values in a preprocessing engine before root-cause isolation. We utilize several imputation methods and compare them in terms of their effectiveness in handling missing syndromes.

The remainder of this chapter is organized as follows. Section 6.1 reviews the background and highlights the contributions of this chapter. Section 6.2 discusses the effects of missing syndromes on a diagnosis system. A number of imputation techniques are used to extend the current diagnosis system to handle missing syndromes. In Sect. 6.3, experimental results on two large-scale data sets, which are generated from complex boards in high-volume production, are used to demonstrate the effectiveness of the proposed diagnosis system in terms of diagnosis accuracy and training time. Finally, Sect. 6.4 concludes the chapter.

## 6.1 Background and Chapter Highlights

Unlike chip-level fault diagnosis, which can be automated via ATE and advanced debug/diagnosis tools [1], board-level diagnosis still requires a considerable amount of manual expertise. Recent work on board-level fault diagnosis has shown that machine learning techniques can be adopted to automate the process of identifying faulty candidates (components) based on the historical data of successfully repaired boards [2, 3].

© Springer International Publishing Switzerland 2017

F. Ye et al., *Knowledge-Driven Board-Level Functional Fault Diagnosis*,
DOI 10.1007/978-3-319-40210-9_6

A diagnostic system based on machine learning does not need to understand the complex functionality of boards, and it is able to automatically derive and exploit knowledge from the repair logs of previously documented cases. The extraction of fault syndromes, i.e., test outcomes, is critical for model training in a diagnosis system. The fault syndromes should provide a complete description of the failure, and the extracted syndromes for different actions should have sufficient diversity such that we can eliminate ambiguity in the eventual repair recommendations. Fault syndromes vary across products and tests. For example, a segment of the log file for a failed traffic test is shown in Fig. 6.1a. The fault syndromes extracted are R3d3–metro (counts mismatch), LA0 Engine 0x0000ffff (error counter), and 0x11 (error code). Each of these elements is considered to be one syndrome. The repair action is often directly recorded in the database, e.g., "replacement of component U11". All the extracted syndromes and actions are used for training of the proposed diagnosis system.

In prior work, e.g., in [3], functional-test logs were parsed in a coarse manner where a syndrome is 1 (*fail*) if a functional test leads to an error at the corresponding observation point. All other functional tests and observation-point pairs, which are mapped to syndromes, are deemed as 0 (*pass*); see Fig. 6.1b [3–5]. However, a syndrome should not be denoted as either *fail* or *pass* when the observation of the test is not available in practice. We propose to define a third syndrome, namely *missing*

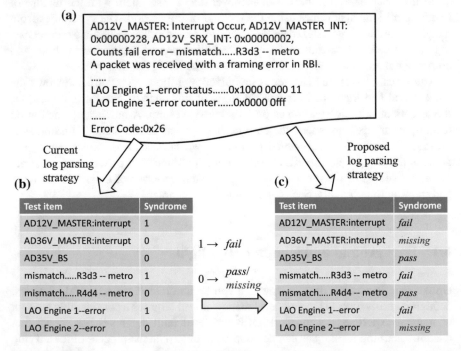

**Fig. 6.1** Illustration of parsing a log with and without *missing* syndromes

syndrome, when the syndrome for a (test, observation point) pair is not observed (see Fig. 6.1c). There are two scenarios that lead to the absence of observation of a syndrome:

(1) **Systematic missing syndromes**: In current test-program design methods, the tasks of designing functional tests are divided into multiple packages and dispatched to multiple design teams [6]. Functional test packages are executed sequentially on a board-under-test. There is a likelihood that the execution of functional tests in later packages is affected by tests in early packages. For example, a functional test, *mainboost*, is used to check the functionality of a global controller, which is required to enable the functional tests that follow [6]. If the controller fails or the attached DRAMs fail, the diagnosis procedure is terminated and subsequent tests cannot be executed. The syndromes corresponding to these tests are, therefore, not available to the diagnosis system.

(2) **Random missing syndromes**: Another scenario in current diagnosis systems is that a few functional tests may occasionally fail to record syndromes. The reason for these occasional syndrome escapes may not be due to board failure, but because of program bugs in functional test design [2, 6]. Such bugs may either lead to incorrectly recorded syndromes or missing observations in the log. In these scenarios, the diagnosis system should be designed to be able to reason in the presence of these missing syndromes.

Both of the above scenarios lead to a decrease in the total number of syndromes available for diagnosis, thus resulting in low diagnosis accuracy. In addition, machine learning-based diagnosis systems proposed in the literature are not equipped to deal with these missing values [4, 7, 8]. These diagnosis systems are not robust to data sets with missing syndromes.

We therefore address the important practical problem of handling missing syndromes in a diagnosis system. The proposed solution preprocesses diagnostic logs based on value imputations; the details are presented in Sect. 6.2. We propose several imputation methods to handle missing values in an effective manner. First, a number of numerical imputation methods are used, such that complete analysis is carried out with the data set of faulty boards, and weighted estimated values are subsequently used to impute missing syndromes. Next, an alternative label imputation method is presented in which an additional feature *missing* is added to explicitly label whether a syndrome is missing.

## 6.2 Methods to Handle Missing Syndromes

Missing data frequently occur in applied data analysis [9, 10]. As described in Sect. 6.1, several scenarios may lead to missing syndromes, e.g., resulting in systematic missing syndromes and random missing syndromes. To address the problem of missing syndromes, a nonconvex max-margin formulation is used in classical

SVM classifiers [10]. However, this method suffers from prohibitively high computation complexity, which makes it infeasible for board diagnosis in realistic scenarios. An alternative method, referred to as *imputation*, is to fill the missing syndromes in the data preprocessing stage [9, 11]. Imputation methods are widely used because of their simplicity and ease of implementation. Imputation based on statistical central tendency (e.g., mode, median, or mean) has often been used in machine learning to treat missing values [11]. The K-nearest neighbor (KNN) and sample mean imputation (SMI) have also been used to carry out the imputation of missing data [12].

## 6.2.1   Missing-Syndrome-Tolerant Fault Diagnosis Flow

A reasoning-based diagnosis flow in [3] can be improved by integrating the component of preprocessing missing values, as shown in Fig. 6.2. A revised diagnosis flow that is capable of handling missing syndromes can be described as follows. Once training data are extracted from diagnosis logs, the diagnosis system determines whether the data set contains missing syndromes, i.e., by comparing the number of obtained syndromes and total number of designed syndromes. If no missing syndrome is detected, the diagnosis system continues with the standard diagnosis system training. Otherwise, the training data set is handled by the missing-value-preprocessing component. The processed data set is then sent for training of the diagnosis system. The same procedure also applies to the new failing boards under diagnosis. The preprocessed new cases are sent to the diagnosis system for root-cause isolation.

The missing-value-preprocessing component contains five parallel subcomponents for handling missing syndromes: (1) complete-case analysis, (2) fractional instances, (3) numerical imputation, (4) label imputation, and (5) feature selection. These subcomponents could either be chosen automatically based on the type of machine learning algorithms used in current diagnosis system, or be set manually by users themselves. Moreover, these subcomponents could even be deliberately combined to better match the characteristics of the current system, leading to higher diagnosis accuracy.

## 6.2.2   Missing-Syndrome-Preprocessing Methods

### 6.2.2.1   Complete-Case Analysis

The basic idea of complete-case analysis is to ignore all the missing values and base the analysis only on complete-case data. Since this method is widely used in the Naive Bayes (NB) classifier, we use a simple NB-based fault diagnosis example to show how this method address missing syndromes.

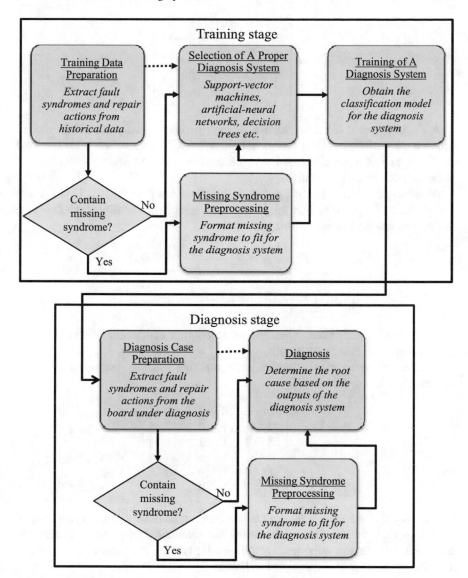

**Fig. 6.2** The SVM-based diagnosis flow with the capability of handling missing syndromes

First, suppose we have a set of faulty boards with two candidate root causes $A_1$ and $A_2$, and we encode them as $y = -1$ and $y = 1$, respectively. Here, we merge the syndromes and the known root causes into a matrix $\mathcal{M} = [\mathcal{B}|\mathcal{C}]$, where the left ($\mathcal{B}$) side refers to syndromes, while the right side ($\mathcal{C}$) refers to the corresponding fault classes. This matrix represents the training information for a diagnosis system that does not contain any missing syndromes,

$$\mathcal{M} = \begin{bmatrix} 1 & 1 & 0 & 1 & \vdots & 1 \\ 0 & 1 & 1 & 0 & \vdots & -1 \\ 0 & 0 & 1 & 1 & \vdots & 1 \\ 0 & 1 & 1 & 1 & \vdots & -1 \\ 1 & 0 & 0 & 0 & \vdots & -1 \\ 1 & 1 & 1 & 0 & \vdots & 1 \end{bmatrix}, \tag{6.1}$$

A Naive Bayes (NB) classifier is a Bayes' theorem-based probabilistic model, which makes inferences from historical data and a prior distribution [13]. The final diagnosis step of a NB classifier is to pick the root-cause candidate with maximal posterior occurrence probability.

Assuming that a new faulty board with syndrome set $T = \{1, 1, 1, 1\}$ is fed into our NB-based diagnosis system, the posterior occurrence probability of this new case can be calculated as shown below:

$$p(A_1|T) = \frac{p(T|A_1) \times p(A_1)}{p(T)} = 0.800$$

$$p(A_2|T) = \frac{p(T|A_2) \times p(A_2)}{p(T)} = 0.200$$

Thus, $A_1$ has higher posterior occurrence probability and would be chosen as the root cause for the new faulty board.

Next, consider the training matrix $\mathcal{M}$ containing missing syndromes, and let ? denotes the missing values:

$$\mathcal{M} = \begin{bmatrix} 1 & 1 & 0 & ? & \vdots & 1 \\ ? & 1 & 1 & 0 & \vdots & -1 \\ 0 & ? & 1 & 1 & \vdots & 1 \\ 0 & 1 & 1 & 1 & \vdots & -1 \\ 1 & 0 & ? & 0 & \vdots & -1 \\ 1 & 1 & 1 & 0 & \vdots & 1 \end{bmatrix} \tag{6.2}$$

The way that the complete-case analysis addresses these missing syndromes is to remove them from training data during the calculation of posterior occurrence probability. Using the same test case $T = \{1, 1, 1, 1\}$, the posterior occurrence probability of this case is calculated as below:

$$p(A_1|T) = \frac{p(T|A_1) \times p(A_1)}{p(T)} = 0.667$$

$$p(A_2|T) = \frac{p(T|A_2) \times p(A_2)}{p(T)} = 0.333$$

The above result shows that $A_1$ is still the most likely root cause for the test case. Thus, in the above example, when the training data contains several missing syndromes, after applying complete-case analysis, our NB classifier can still give accurate fault diagnosis.

Complete-case analysis can also be used with other machine learning models such as SVM and ANN. However, complete-case analysis in these models will discard the entire instances containing missing values instead of just ignoring the missing values themselves, resulting in much lower diagnosis accuracy than the NB classifier when the amount of missing information is relatively high.

### 6.2.2.2  Fractional Instances

The goal of the fractional-instances method is to distribute missing syndromes over other complete instances. An instance containing missing values will be split into multiple weighted instances. Each weighted instance represents one possible value combination of those missing syndromes. The weight value associated with each split instance is the fraction of complete cases that have the same syndrome values as this split instance.

Let us assume that we apply this fractional-instance method to the training matrix with missing syndromes shown in (6.2). A new matrix is obtained as:

$$
\mathscr{M} =
\begin{bmatrix}
0.4 & 0.4 & 0 & 0.4 & \vdots & 1 \\
0.6 & 0.6 & 0 & 0 & \vdots & 1 \\
0.6 & 0.6 & 0.6 & 0 & \vdots & -1 \\
0 & 0.4 & 0.4 & 0 & \vdots & -1 \\
0 & 0.8 & 0.8 & 0.8 & \vdots & 1 \\
0 & 0 & 0.2 & 0.2 & \vdots & 1 \\
0 & 1 & 1 & 1 & \vdots & -1 \\
0.8 & 0 & 0.8 & 0 & \vdots & -1 \\
0.2 & 0 & 0 & 0 & \vdots & -1 \\
1 & 1 & 1 & 0 & \vdots & 1
\end{bmatrix}
\tag{6.3}
$$

The advantage of the fractional-instances method is that it greatly enriches the incomplete training data by artificially adding instances to represent every possible value combinations of missing syndromes.

However, the number of additional instances will increase exponentially with the number of missing values, making this method infeasible for machine learning algorithms that require training data using the entire syndrome set, e.g., traditional ANNs and SVMs. In contrast, the decision tree (DT), which is a tree-like predictive model for mapping observations of an item to its targeted value [14], is suitable for this method since a DT typically only needs a subset of syndromes for training.

### 6.2.2.3   Numerical Imputation

Numerical imputation is widely used to infer missing values. A *missing* value is predicted on a numerical scale based on analysis of the failing boards. Several imputation methods, e.g., mean and mode, can be used [9, 11] and they are described in detail later. Suppose that we have the same training matrix with missing syndromes as shown in (6.2). After applying numerical imputation, the new matrix is given by

$$\mathcal{M} = \begin{bmatrix} 1 & 1 & 0 & b_{41} & \vdots & 1 \\ b_{12} & 1 & 1 & 0 & \vdots & -1 \\ 0 & b_{23} & 1 & 1 & \vdots & 1 \\ 0 & 1 & 1 & 1 & \vdots & -1 \\ 1 & 0 & b_{35} & 0 & \vdots & -1 \\ 1 & 1 & 1 & 0 & \vdots & 1 \end{bmatrix} \tag{6.4}$$

where $b_{ij}$ is the missing value of syndrome $t_i$ collected from the failing board $j$. Different imputation methods can be used to assign values for $b_{ij}$; these are discussed below.

(I)  Zero Imputation: In the prior log parsing strategy in Fig. 7.3b, all *missing* syndromes are deemed to be *pass*, which is denoted as 0. Zero imputation follows the same imputation method, where any missing syndrome $b_{ij}$ is imputed with 0. As an illustration, (6.4) can be revised with zero-imputation values as

$$\mathcal{M} = \begin{bmatrix} 1 & 1 & 0 & \mathit{0} & \vdots & 1 \\ \mathit{0} & 1 & 1 & 0 & \vdots & -1 \\ 0 & \mathit{0} & 1 & 1 & \vdots & 1 \\ 0 & 1 & 1 & 1 & \vdots & -1 \\ 1 & 0 & \mathit{0} & 0 & \vdots & -1 \\ 1 & 1 & 1 & 0 & \vdots & 1 \end{bmatrix}, \tag{6.5}$$

where italics indicate the imputed values and are used to denote imputed values throughout this paper.

This type of a missing-syndrome-handling method is widely used in the artificial neural network (ANN) model, which is a supervised machine learning method that has also been widely used for pattern classification and related problems [15, 16].

(II)  Mode-Value Imputation: Assume that $A_i$ is the full set of entries for syndrome $x_i$, $A_i'$ is the subset of $A_i$ that contains all the nonmissing entries, where $a_{ij} \neq missing$, and $a_{ij} \in A_i$. In the mode-value imputation method, we use the following equation to impute missing values.

$$b_{ij} = mode(A_i') = \begin{cases} 1, & \forall\, P(a_{ij} = 1, a_{ij} \in A_i') > 0.5 \\ 0, & \forall\, P(a_{ij} = 1, a_{ij} \in A_i') \leq 0.5, \end{cases}$$

where the imputation value $b_{ij}$ is the same for each syndrome. As an illustration, the corresponding synthetic data in (6.4) is revised as:

$$\mathcal{M} = \begin{bmatrix} 1 & 1 & 0 & 0 & \vdots & 1 \\ 1 & 1 & 1 & 0 & \vdots & -1 \\ 0 & 1 & 1 & 1 & \vdots & 1 \\ 0 & 1 & 1 & 1 & \vdots & -1 \\ 1 & 0 & 1 & 0 & \vdots & -1 \\ 1 & 1 & 1 & 0 & \vdots & 1 \end{bmatrix} \tag{6.6}$$

Mode imputation may be ineffective when the probability of the fault syndrome being '0' or '1' are similar. This method will always impute the value with higher occurrence probability, which may distort the original distribution.

(III) Mean-Value Imputation: Assume that $A_i$ is the complete set of entries for syndrome $x_i$, and $A_i'$ is the subset of $A_i$ that contains all the nonmissing entries, where $a_{ij} \neq missing$ and $a_{ij} \in A_i$. In this method, a mean value $b_{ij}$ is computed based on $A_i'$, such that the imputation value $b_{ij}$ can be regarded as the probability that the syndrome $x_i$ is *fail*. The imputation value $b_{ij}$ is calculated as

$$b_{ij} = mean(A_i') = \frac{\sum_{a_{ij} \in A_i'} a_{ij}}{|A_i'|}, \tag{6.7}$$

where the imputation value $b_{ij}$ is same for each syndrome $x_i$. As an illustration, the corresponding matrix for the synthetic data is updated as:

$$\mathcal{M} = \begin{bmatrix} 1 & 1 & 0 & 0.4 & \vdots & 1 \\ 0.6 & 1 & 1 & 0 & \vdots & -1 \\ 0 & 0.8 & 1 & 1 & \vdots & 1 \\ 0 & 1 & 1 & 1 & \vdots & -1 \\ 1 & 0 & 0.8 & 0 & \vdots & -1 \\ 1 & 1 & 1 & 0 & \vdots & 1 \end{bmatrix} \tag{6.8}$$

The combination of mean and mode imputation is widely used in the Support-Vector Machine (SVM) model, which is a supervised machine learning technique [17] and its goal is to define an optimal separating hyperplane (OSH) to separate two classes. Generalizations also exist for handling more than two classes. For features having continuous values, the SVM's default mechanism inserts mean values into missing positions while for features with discrete values, it imputes mode values to replace missing syndromes.

### 6.2.2.4 Label Imputation

The numerical format of syndromes shown in Fig. 7.3b is no longer sufficient to describe the observations in the log of functional tests when we introduce a third syndrome. Therefore, we propose to use another label, referred to as *missing*, to denote the absence of a syndrome;

$$
\mathcal{M} =
\begin{bmatrix}
pass & pass & fail & missing & 1 \\
missing & pass & pass & fail & -1 \\
fail & missing & pass & pass & 1 \\
fail & pass & pass & pass & -1 \\
pass & fail & missing & fail & -1 \\
pass & pass & pass & fail & 1
\end{bmatrix}
\tag{6.9}
$$

see Fig. 7.3c. Equation (6.4) can now be updated as Eq. (6.9).

Machine learning techniques can handle label syndromes by converting labels to numerical values [9]. For a ternary syndrome, two values are usually used to represent one syndrome. The first value $a_{ij}$ still denotes the failure observed in a syndrome, such that it is 1 if a syndrome is observed as *fail*, and 0 otherwise. The second value $a'_{ij}$ denotes the occurrence of a syndrome, such that a missing syndrome is denoted with a weighted value $d$, and 0 otherwise. The reason for using these weighted values is that the additional feature $a'_{ij}$ may mask the relationship between the original syndrome-fault pair. Fault syndromes are usually the dominant features for root-cause isolation, while missing syndromes are the secondary features. Therefore, it is necessary to select an appropriate value $d$ in label imputation. The updated synthetic board is as follows:

$$
\mathcal{M} =
\begin{bmatrix}
1 & (0) & 1 & (0) & 0 & (0) & 0 & (d_{41}) & 1 \\
0 & (d_{12}) & 1 & (0) & 1 & (0) & 0 & (0) & -1 \\
0 & (0) & 0 & (d_{23}) & 1 & (0) & 1 & (0) & 1 \\
0 & (0) & 1 & (0) & 1 & (0) & 1 & (0) & -1 \\
1 & (0) & 0 & (0) & 0 & (d_{35}) & 0 & (0) & -1 \\
1 & (0) & 1 & (0) & 1 & (0) & 0 & (0) & 1
\end{bmatrix}
\tag{6.10}
$$

where $d_{ij}$ is the weighted value for the $j$th fault syndrome of the $i$th instance. We use the following weighted mean-value imputation to calculate $d_{ij}$,

$$
d_{ij} = \frac{mean(A'_j)}{w} = \frac{\sum_{a_{ij} \in A'_i} a_{ij}}{w|A'_i|},
\tag{6.11}
$$

where $w$ is a single weight for tuning the feature of missing syndromes, and it is the same for imputing all syndromes. Choosing an appropriate value of $w$ is important for a diagnosis system in order to achieve high diagnosis accuracy. We provide guidelines for selecting $w$ in Sect. 6.3.

## 6.2.3   Feature Selection

Feature selection, also referred as subset selection, is used to select an effective, but reduced, set of syndromes for use by a diagnosis system. The main idea of methods for handling missing syndromes is to add statistical information to the incomplete data set to compensate for losses caused by missing syndromes. However, the additional information may also introduce irrelevant, redundant, or even misleading syndromes, lowering the diagnosis accuracy of the original system.

Therefore, the goal of feature selection is to identify a set of most important features from incomplete data for characterization. One of the most popular solutions for the subset-selection problem is based on the metric of minimum-redundancy-maximum-relevance (*mRMR*). Details of *mRMR* method was described in Chap. 5. We adapt *mRMR* subset selection process for handling missing syndromes. Two approaches are discussed below.

### 6.2.3.1   Complete-Case Analysis

The first method that we use to address missing syndromes in $mRMR$ subset selection is the complete-case analysis as described in Sect. 6.2.2.1. Suppose we have a set of successfully repaired faulty boards with root cause set $\mathbf{A} = \{A_1, A_2, \ldots, A_N\}$ and syndrome set $\mathbf{T} = \{T_1, T_2, \ldots, T_M\}$. We can compute the desired posterior occurrence probability of root cause $A_j$, $p(A_j|T)$. Then, the feature selection approach can calculate the relevance values, redundancy values, and $mRMR$ values, using this set of posterior. The final $mRMR$ syndrome subset can then be determined by selecting syndromes with largest $mRMR$ values.

### 6.2.3.2   Label Imputation

The second approach we use here is to introduce label syndromes such that each missing syndrome can be treated as a separate value. A simple example can help better illustrate how this method works. In Eq. (6.12), the $\rightarrow$ indicates that after feature selection, the original matrix on the left side has been reduced to the matrix on the right side. We can see that only syndrome $T_1$ is preserved in the new syndrome set. Now, assume we have several missing values in the original matrix and we would like to treat them as separate values during feature selection. The new process is shown in Eq. (6.13). We can see that label imputation doubles the size of the syndrome set, providing more information for the subsequent subset-selection step. In the final reduced-syndrome set, both $T_1$ and $T_2$ are preserved for future diagnosis while $T_3$ and other irrelevant label syndromes are removed after feature selection.

$$\mathcal{M} = \begin{bmatrix} 1 & 0 & 1 & 1 \\ 0 & 1 & 0 & -1 \\ 1 & 0 & 0 & 1 \\ 0 & 1 & 0 & -1 \end{bmatrix} \rightarrow \begin{bmatrix} 1 & 1 \\ 0 & -1 \\ 1 & 1 \\ 0 & -1 \end{bmatrix} \tag{6.12}$$

$$\mathcal{M} = \begin{bmatrix} 1 & ? & 1 & 1 \\ ? & 1 & ? & -1 \\ 1 & ? & ? & 1 \\ ? & 1 & 0 & -1 \end{bmatrix} \rightarrow \begin{bmatrix} 1 & 0 & ? & d_{21} & 1 & 0 & 1 \\ ? & d_{12} & 1 & 0 & ? & d_{32} & -1 \\ 1 & 0 & ? & d_{23} & ? & d_{33} & 1 \\ ? & d_{14} & 1 & 0 & 0 & 0 & -1 \end{bmatrix} \rightarrow \begin{bmatrix} 1 & 0 & 1 \\ 0 & 1 & -1 \\ 1 & 0 & 1 \\ 0 & 1 & -1 \end{bmatrix} \tag{6.13}$$

The comparison between these two methods in terms of diagnosis accuracy and subset size is described in Sect. 6.3.

## 6.3   Results

Experiments were performed on one synthetic board and one industrial board. Both boards under study are line process units (LPUs) in a high-end router with 10-Gbit/s interfaces, which are designed for core and backbone commercial networks. Table 6.1 presents some basic information about the two boards used in our experiments, including number of syndromes, number of root causes, and number of instances.

To evaluate the performance of the proposed methods for handling missing syndromes, we need to generate missing syndromes in our training data. The occurrence of missing syndromes in current diagnosis systems is most likely due to program bugs in the interface of gathering data from test equipment instead of board failures or test-design incompleteness. Such bugs may either lead to incorrectly recorded syndromes or missing observations in the log files. Since missing syndromes of all types (and anywhere in the system response) can result due to these data-gathering-interface bugs, we assume that missing syndromes can occur in any position in our training data. Let $T$ be the entire set of training set, $S$ be the total number of syndromes, and $C$ be the total number of failing boards. Let $X$ be the total number of missing syndromes in $T$. The missing ratio (MR) is defined as $\frac{X}{S \times C}$. In this work, we use four values for the missing ratio, 10, 30, 50, and 70%, to consider a wide range of scenarios. Since most machine learning methods employ heuristics and the

**Table 6.1**   Information about the boards used for classification

|                        | Board 1 | Board 2 |
|------------------------|---------|---------|
| Number of syndromes    | 207     | 898     |
| Number of root causes  | 14      | 222     |
| Number of boards       | 1400    | 2965    |

missing syndromes can be randomly distributed, Monte Carlo simulation is necessary to ensure that our results are not favorable as a result of serendipity. We carried out 500 Monte Carlo simulation runs because there was no significant change in the results when we increased this number to over 500. For less than 500 Monte Carlo runs, we observed significant variation in the results. The average success ratios are presented to show the diagnosis performance with different missing-syndrome-handling methods.

All the algorithms are implemented in an open-source Machine learning toolkit, namely WEKA [18]. Experiments were run on a 64-bit Linux system with 12 GB of RAM and quadcore Intel i7 processors running at 2.67 GHz.

To evaluate the diagnosis performance of different machine learning systems, we use a five-fold *cross-validation* method, which randomly partitions the training set into five groups. Each group is regarded as a test case while all of the other cases are fed for training. The success ratio (SR), referred to as a percentage, is used to measure the diagnosis accuracy. In addition, since SR is a coarse metric for multiclass classifiers (e.g., a board-level functional fault diagnosis system) and does not provide any suggestion for improving diagnosis, we will also use a fine-grained set of information-theoretic metrics, called *precision* and *recall*, to comprehensively evaluate the diagnosis system. The details of *precision* and *recall* are described in Chap. 5. In board-level diagnosis, *precision* describes the percentage of success in predicting a root cause, while *recall* reflects the percentage of success for a root cause to be predicted. A combination of these two metrics provides a more complete picture.

### 6.3.1 Evaluation of Label Imputation

First, we evaluate the label imputation method. Different weight values, $1 \leq w \leq 20$, in label imputation are compared in terms of success ratios in diagnosis. Similar trends of weight values have been obtained for both boards under study. We present the results of Board 1 here, as shown in Fig. 6.3. We observe that although the trends of weight values depend on type of machine learning models as well as missing ratios, too small or large a choice of $w$ typically leads to reduced diagnosis accuracy. For example, for the SVM model in Board 1, if we consider 50 % missing syndromes, the SR is only 75 % for $w = 1$. As we increase $w$, the SR increases to 91 % for $w = 5$. However, as we continue to increase $w$, the SR slightly decreases. The SR obtained by using $w = 20$ is 88 %.

Similar SR results are obtained for other missing ratios and machine learning algorithms for Board 1 in Fig. 6.3. The reason smaller $w$ leads to lower SR is that the use of missing syndromes with a small value of $w$ masks the fail syndromes; the fail syndromes are, however, more informative for diagnosis. In contrast, the reason that too large a value of $w$ leads to lower SR is that the missing syndromes do not contribute to diagnosis in these cases. If $w$ is infinite, the imputation value $d$ becomes 0 for all missing syndromes.

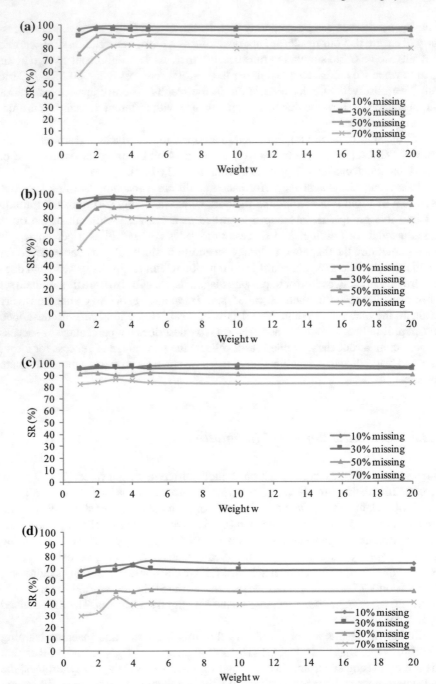

**Fig. 6.3** Comparison of SR using label imputation with different weights for Board 1. **a** SVM. **b** ANN. **c** Bayes. **d** DT

### 6.3.2 Evaluation of Feature Selection in Handling Missing Syndromes

We compare the two feature selection methods in terms of the size of the reduced-syndrome set they provide. Since both boards show similar trends in the size of extracted subset, we present the results for Board 1 and Board 2 in Fig. 6.4; In Fig. 6.4, "Subset_M1" refers to the use of complete-case analysis to deal with missing syndromes during feature selection while "Subset_M2" refers to the use of label imputation to address missing values. First, we can see that for both M1 and M2, with an increase in the missing ratio, the size of the extracted syndrome set after feature selection increases first, decreases later, and eventually converges. One possible reason for this phenomenon is that since feature selection is used to extract a set of most informative features for a given board, when the logs contain missing

**Fig. 6.4** Subset size of two feature-selection-based methods. **a** Board 1. **b** Board 2

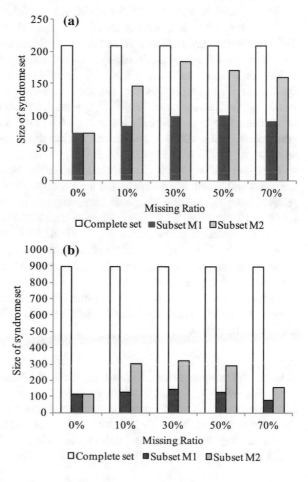

syndromes, then some original informative features may no longer provide useful information; thus feature selection may have to include more alternative features in its extracted subset so that this extracted subset can still give satisfactory diagnosis accuracy. However, feature selection cannot find more appropriate alternative syndromes when the missing ratio is too high. Second, we can see that M2 preserves more syndromes than M1. This is because M2 applies label imputation to deliberately add extra information for missing syndromes while M1 only discards missing syndromes.

## 6.3.3  Comparison of Different Missing-Syndrome Handling Methods

In Figs. 6.5 and 6.6, we apply various missing-syndrome handling methods to four different machine learning models: SVM, ANN, Naive Bayes, and Decision Tree, and then compare their diagnosis accuracies obtained for Board 1 and Board 2 under different missing ratios. Note that the "default missing handling methods" in Figs. 6.5 and 6.6 refers to the default missing-syndrome handling method in these four learning models in WEKA as shown in Table 6.2. In addition, the "feature selection_M1" is the feature selection process that uses complete-case analysis to deal with missing values while "feature selection_M2" applies label imputation to address missing syndromes. The dashed line shown in Figs. 6.5 and 6.6 is the diagnosis accuracy of zero missing ratio, which can be seen as our baseline. The results can be summarized as follows:

First, there is a significant diagnosis accuracy gap between the synthetic board and the industrial board. For example, for the four machine learning models, i.e., SVM, ANN, Naive Bayes, and Decision Tree, the success ratios for 0% missing ratio for Board 1 are 99.12%, 98.89%, 98.79% and 82.67%, respectively. For Board 2, the baseline success ratios are 78.76, 77.97, 81.67, and 70.56%. In addition, from Figs. 6.5 and 6.6, we can see that with an increase of missing ratio, the diagnosis accuracy of synthetic boards and industrial boards exhibit a similar decreasing trend for different machine learning models.

Second, various missing-syndrome-handling methods perform significantly differently for different machine learning models under different missing ratios. Details are discussed below.

### 6.3.3.1  Missing-Syndrome Handling for SVM-Based Diagnosis System

Since syndromes in our diagnosis system are denoted as nominal values '0' and '1', the default method for an SVM in our experiments is mode imputation. The result shows that the label imputation method always leads to higher diagnosis accuracies for Board 1 (Fig. 6.5a) and Board 2 (Fig. 6.6a). The two feature-selection-based

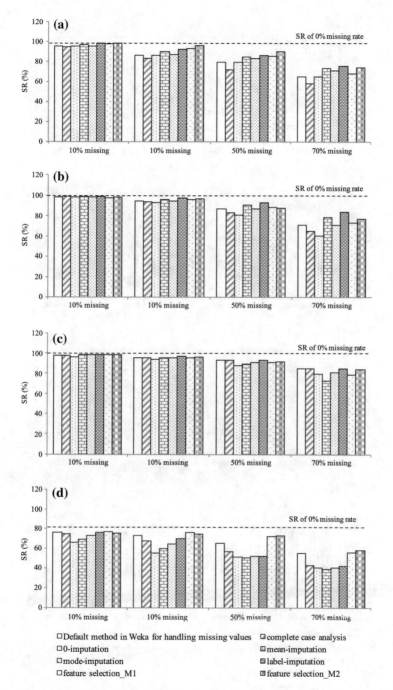

**Fig. 6.5** Comparison of success ratios for diagnosis using different missing-syndrome handling methods for Board 1. **a** SVM. **b** ANN. **c** Bayes. **d** DT

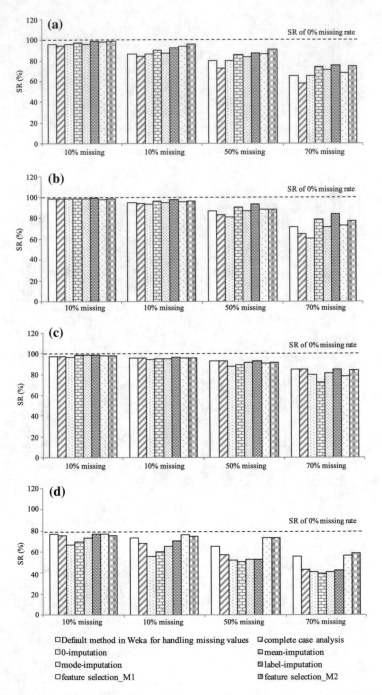

**Fig. 6.6** Comparison of success ratios for diagnosis using different missing-syndrome handling methods for Board 2. **a** SVM. **b** ANN. **c** Bayes. **d** DT

**Table 6.2** The default technique for handling missing syndromes in the WEKA machine learning package

| | Default method for handling missing syndromes |
|---|---|
| SVM | Mode imputation (discrete features) |
| | Mean imputation (continuous features) |
| ANN | Zero imputation |
| Naive Bayes | Complete-case analysis |
| Decision Tree | Fractional instances |

methods also perform well, with diagnosis accuracy comparable to mean imputation under different missing ratios. For example, if we consider a 50% missing ratio, the average SRs for Board 1 are 93.42, 90.79, 88.39, 88.24, 86.84, and 83.47, 80.42% for label imputation, mean imputation, feature selection_M1, feature selection_M2, mode imputation (default in WEKA), complete-case analysis, and zero imputation, respectively.

#### 6.3.3.2 Missing-Syndrome Handling for ANN-Based Diagnosis System

Zero imputation is the default method in ANN's WEKA implementation. From Figs. 6.5b and 6.6b, we can see that although label imputation is effective, feature selection_M2, which applies label imputation to its feature selection process, performs even better. Taking a 50% missing ratio of Board 1 as an example, feature selection_M2 achieves 89.56% diagnosis accuracy, about 14% higher than the default mechanism and 5% higher than label imputation.

#### 6.3.3.3 Missing-Syndrome Handling for NB-Based Diagnosis System

Complete-case analysis is the default method that WEKA uses with its Nave Bayes classifier. Unlike in other learning models, this method yields impressive results for different missing ratios. Still using the 50% missing ratio case of Board 1 in Fig. 6.5c as our example, the difference between the 93.03% SR of label imputation and the 93.16% SR for complete-case analysis is negligible.

#### 6.3.3.4 Missing-Syndrome Handling for DT-Based Diagnosis System

The decision tree model in WEKA, J48, incorporates the fractional instances method to address missing syndromes. As shown in Figs. 6.5d and 6.6d, the most interesting observation here is that all imputation methods (including label imputation) "fail" when the missing ratio is over 50%. In contrast, the two feature-selection-based methods are still effective, and provide better results when the missing ratio is high. For the 50% missing ratio, the diagnosis accuracy of all imputation methods fall to about 50% SR while the feature selection M1 and M2 still maintain their success ratio over 70%.

In addition to the success ratio, we also utilize the precision and recall criteria to carry out more fine-grained analysis. We investigate the effect of missing syndromes on root-cause isolation. We present the results of Board 1 and Board 2 in Figs. 6.7 and 6.8. The results show that when the missing ratio is low, all four machine learning techniques can identify most root causes without ambiguity, but when the missing ratio is higher, root causes cannot be clearly differentiated from each others. For example, in Fig. 6.7, when only 10% of the values are missing, most root causes have precision and recall over 0.8. In contrast, when the missing ratio increases to 70%, the precision and recall of most root causes are distributed between 0.4 and 0.7, dropping significantly from the 10% missing ratio. For the Board 2, Fig. 6.8 present its precision and recall under different missing ratios. Similar to the synthetic boards, the number of root causes that have high precision and recall decreases significantly with the increasing of missing ratios. For example, Fig. 6.8d shows that for the DT model, when the missing ratio is 70%, 115 out of 222 root causes have both precision and recall lower than 0.3, which means over half of root causes have become undistinguished.

## 6.3.4    Evaluation of Training Time

Finally, we compare the computational complexity of label imputation and the two feature-selection-based methods in terms of diagnosis system training time. Although the training time for Board 2 are much higher than Board 1, both boards have similar trends of training time for different missing-syndrome-handling methods. The result of Board 1 is presented in Fig. 6.9. The results show that the training time for label imputation is much higher than that for the other methods. For Board 1, the training time of SVM and DT model is an average of 40 min, and the ANN model requires over 2 h. In contrast, the two proposed feature-selection-based methods need only a few minutes for SVM, Bayes, and the DT models, and 50 min for the ANN model. The training time of the Board 2 is much higher than that of synthetic Board 1. However, the two proposed feature-selection-based methods still need less training time than other methods. Therefore, the two feature-selection-based methods are also computationally more efficient.

In summary, in our experiments with a synthetic board and an industrial board, the method "feature-selection-M2" performs better in handling missing syndromes in terms of diagnosis accuracy and training time for most machine learning models. However, since different boards may have significantly different characteristics, we should not conclude that "feature-selection-M2" will always yield the best results. A

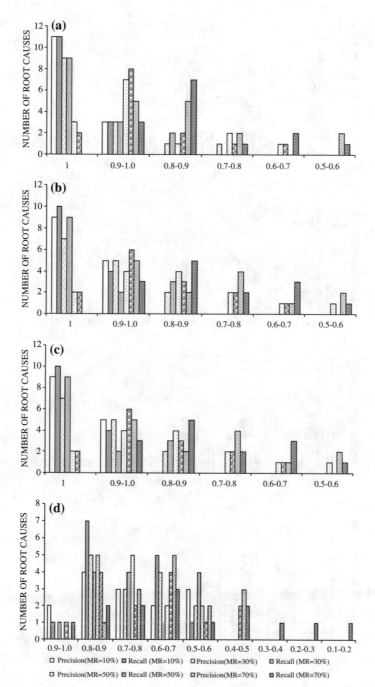

**Fig. 6.7** The precision and recall under various missing ratios for Board 1. **a** SVM. **b** ANN. **c** Bayes. **d** DT

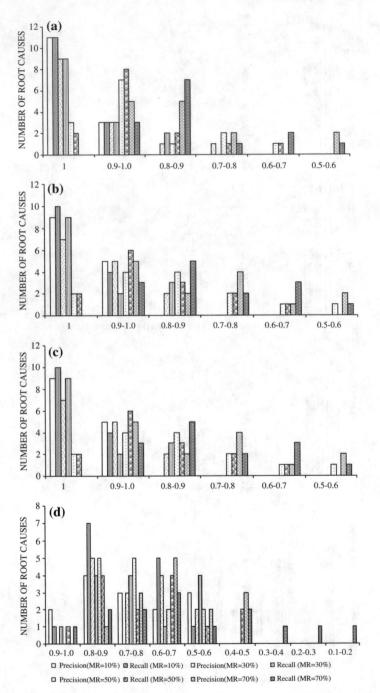

**Fig. 6.8** The precision and recall under various missing ratios for Board 2. **a** SVM. **b** ANN. **c** Bayes. **d** DT

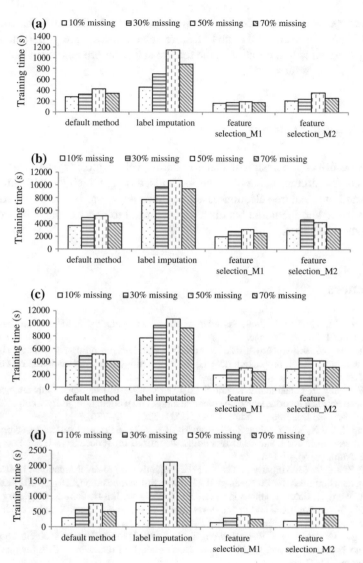

**Fig. 6.9** Comparison of training time for Board 1. **a** SVM. **b** ANN. **c** Bayes. **d** DT

more realistic conclusion is that the choice of methods to handle missing syndromes depends on two factors:

(1) Which machine learning method is used: Different machine learning models may prefer different missing-syndrome-handling methods. For example, although label imputation performs best in the ANN model, it "fails" in the DT model.
(2) Which objectives are considered: Different fault diagnosis system may have different objectives. For example, the goal may be to not only have high diagnosis

accuracy, but also to reduce irrelevant syndromes. In this case, the two methods "feature-selection-M1" and "feature-selection-M2" are preferable because they have already incorporated the feature selection process when dealing with missing syndromes.

## 6.4   Chapter Summary

We have described the design of a smart board-level diagnosis system that can handle missing syndromes using different methods, among which the label imputation method and two feature-selection-based methods appear to be the most promising. Both synthetic and industrial boards have been used to validate the effectiveness of the proposed methods.

## References

1. Bushnell M, Agrawal V (2000) Essentials of electronic testing for digital, memory, and mixed-signal VLSI circuits. Springer, New York
2. O'Farrill C, Moakil-Chbany M, Eklow B (2005) Optimized reasoning-based diagnosis for non-random, board-level, production defects. In: Proceedings IEEE international test conference (ITC), pp 173–179
3. Ye F, Zhang Z, Chakrabarty K, Gu X (2013) Board-level functional fault diagnosis using artificial neural networks, support-vector machines, and weighted-majority voting. IEEE Trans Comput-Aided Des Integr Circuits Syst (TCAD) 32(5):723–736
4. Zhang Z, Gu X, Xie Y, Wang Z, Chakrabarty K (2012) Diagnostic system based on support-vector machines for board-level functional diagnosis. In: Proceedings IEEE European test symposium (ETS), pp 1–6
5. Ye F, Zhang Z, Chakrabarty K, Gu X (2012) Adaptive board-level functional fault diagnosis using decision trees. In: Proceedings IEEE Asian test symposium (ATS), pp 202–207
6. Vo T, Wang Z, Eaton T, Ghosh P, Li H, Lee Y, Wang W, Jun H, Fang R, Singletary D, Gu X (2006) Design for board and system level structural test and diagnosis. In: Proceedings IEEE international test conference (ITC), pp 1–10
7. Zhang Z, Chakrabarty K, Wang Z, Wang Z, Gu X (2011) Smart diagnosis: efficient board-level diagnosis and repair using artificial neural networks. In: Proceedings IEEE international test conference (ITC), pp 1–10
8. Amati L (2012) Test and diagnosis strategies for digital devices: methodologies and tools. PhD dissertation, Politecnico di Milano, Italy
9. Pelckmans K et al (2005) Handling missing values in support vector machine classifiers. Neural Netw 18(5):684–692
10. Chechik G et al (2008) Max-margin classification of data with absent features. J Mach Learn Res 9:1–21
11. Saar-Tsechansky M, Provost F (2007) Handling missing values when applying classification models. J Mach Learn Res 8:1625–1657
12. Hruschka ER et al (2003) Evaluating a nearest-neighbor method to substitute continuous missing values. Advances in artificial intelligence. Springer, Berlin, pp 723–734
13. Rish I (2001) An empirical study of the naive bayes classifier. IBM New York workshop on empirical methods in artificial intelligence, vol 3(22), pp 41–46

14. Quinlan J (1986) Induction of decision trees. Mach Learn 1(1):81–106
15. Ding Y, Peng X, Fu X (2009) The research of artificial neural network on negative correlation learning. Advances in computation and intelligence. Springer, New York, pp 392–399
16. Prieto B, de Lope J, Maravall D (2005) Reconfigurable hardware implementation of neural networks for humanoid locomotion. Artificial intelligence and knowledge engineering applications: a bioinspired approach. Springer, New York, pp 395–404
17. Vapnik V (1995) The nature of statistical learning theory. Springer, New York
18. Chang C-C, Lin C-J (2011) LIBSVM: a library for support vector machines. ACM transactions on intelligent systems and technology, vol 2, pp 27:1–27:27. Software available at http://www.csie.ntu.edu.tw/~cjlin/libsvm

# Chapter 7
# Knowledge Discovery and Knowledge Transfer

Reasoning-based methods have recently become popular since they overcome the knowledge-acquisition bottleneck during volume production and they can automatically generate an intelligent diagnostic system from existing resources [1, 2]. However, knowledge acquisition is a major problem for a reasoning-based method at the initial product ramp-up stage. Machine learning-based reasoning requires an adequate database for training the reasoning engine, and such a database becomes available much later in the product cycle.

In this chapter, we propose a knowledge-discovery method and a knowledge-transfer method for facilitating board-level functional fault diagnosis at the initial product ramp-up stage. The proposed methods help address the knowledge gap between test design stage and volume production. First, an analysis technique based on topic model is used to discover knowledge from syndromes, which can be used for training a diagnosis engine. Second, knowledge from diagnosis engines used for earlier-generation products can be automatically transferred through root-cause mapping and syndrome mapping based on keywords and board-structure similarities.

The remainder of this chapter is organized as follows. Section 7.1 describes the "data acquisition" problem in detail and highlights the contribution of this work. Section 7.2 presents an overview of the knowledge discovery and transfer framework in reasoning-based fault diagnosis. Details of the knowledge-discovery solution are described in Sect. 7.3 and the knowledge-transfer mechanism is described in Sect. 7.4. Results on knowledge discovery and transfer involving five boards from industry are presented in Sect. 7.5. The high diagnosis accuracy obtained using the new diagnosis engine demonstrates the effectiveness of the proposed knowledge discovery and transfer framework. Finally, Sect. 7.6 concludes the chapter.

## 7.1 Background and Chapter Highlights

Reasoning-based diagnostic systems based on machine learning can automatically derive and exploit knowledge from repair logs of previously documented cases, without requiring detailed understanding of the complex functionality of boards

© Springer International Publishing Switzerland 2017
F. Ye et al., *Knowledge Driven Board-Level Functional Fault Diagnosis*,
DOI 10.1007/978-3-319-40210-9_7

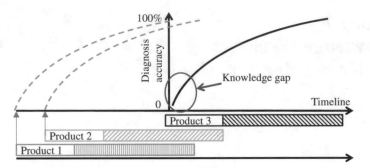

**Fig. 7.1** The problem of low accuracy for a reasoning-based diagnosis system at the beginning of a new product (Product 3) ramp-up stage

[3, 4]. The diagnosis accuracy improves when more successfully repaired boards during volume production are available for training the diagnosis system by using incremental learning [5, 6]. However, there exists a significant knowledge gap in the initial product ramp-up stage, even though knowledge accumulation is badly needed for diagnosis system training, as shown in Fig. 7.1. Knowledge acquisition is the major bottleneck for developing a usable diagnosis engine. In contrast, rule-based and model-based techniques [7, 8] can assist in the initial creation of the knowledge base, but the acquisition of such knowledge requires a technician's expertise and the knowledge may be biased or flawed if the board structure is not fully understood. Model-based reasoning [9] is a hybrid model/reasoning-based approach to bridge the knowledge gap between initial ramp-up and volume production. However this approach is deployed mainly in the form of a model-based method, and reasoning is used only in a limited way; thus considerable human effort is required to maintain the diagnosis engine. The knowledge gap results in ambiguous root-cause prediction, which in turn leads to longer diagnosis time, low yield, and high manufacturing cost.

To view this problem in a more common and realistic scenario, a successful product typically experiences multiple updates and there are often similar products during a period of time; see Fig. 7.1. From the hardware point of view, each version update involves incremental changes to the board structure, and the corresponding functional test program needs minor changes as well. The test programs are continually updated until the product becomes mature. However, even if the components on a board and the test programs change incrementally, traditional diagnosis engine has to be trained using the cases diagnosed with the new test programs. Due to the low volume in the ramp-up stage, the number of failing boards for training the new diagnosis engine is very small at this stage. Therefore, it is infeasible to deploy learning-based diagnosis engine in the early manufacturing phase of a new product. Hence, there is a major incentive to explore new methods of knowledge acquisition. We propose a knowledge-discovery and knowledge-transfer framework as a solution to the problem of knowledge acquisition for the newer-generation diagnosis engine.

The contributions of this work can therefore be detailed as follows:

1. Automated knowledge acquisition and filling of the knowledge gap during the product ramp-up stage.
2. Bridging the model-based diagnosis engine with reasoning-based diagnosis to overcome the knowledge acquisition bottleneck at the ramp-up stage.
3. Automated extraction of the relationship between root causes and syndromes through the concept of topic model.
4. Knowledge discovery and reuse of prior knowledge from the previous diagnosis system and its use in the new diagnosis engine, thus enabling knowledge inheritance among products.

## 7.2 Overview of Knowledge Discovery and Transfer Framework

The proposed knowledge-discovery and knowledge-transfer framework enhances the capability of a reasoning-based diagnosis engine, by providing knowledge from *insourcing* and *outsourcing* in the form of synthetic cases using the proposed knowledge discovery and knowledge-transfer methods, respectively, as shown in Fig. 7.2.

*Insourcing* relies on knowledge discovery, which extracts knowledge, reflected as the relationship between root causes and test programs, from the test logs. In prior work [3], syndromes extracted from the test log and the recorded root-causes are fed to the diagnosis engine for learning. Defining and extracting syndromes from logs is a critical step in the development of an automatic diagnosis system. However, this step relies on the experience of test engineers and hence it consumes a significant amount of time. Once syndromes are defined, they are simply features for the classification engine. Not much attention has been paid in the past to the content of these syndromes and the automation of syndrome extraction, even though syndromes include valuable information for root-cause isolation. The knowledge-

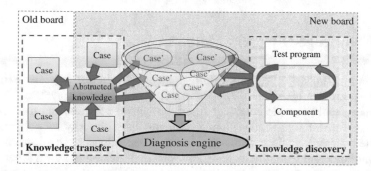

**Fig. 7.2** Illustration of knowledge-discovery and transfer framework for board-level diagnosis

discovery approach can facilitate engineers to mine useful knowledge from the log, reflected as specific keywords, because the test program is designed intentionally to contain as much (unstructured) information as possible to help isolate root-causes. The proposed knowledge-discovery approach can be regarded as an example of a "topic model", which is discussed in detail in Sect. 7.3.

*Outsourcing* relies on knowledge transfer. "Learning" in prior work on the design of reasoning-based diagnosis engine can be viewed as direct self-learning [1, 3]. The diagnosis engine acquires knowledge from the failing boards of its own board-type. Self-learning is useful when boards are in volume production, but it is less effective when an adequate number of failing boards is not available to learn from at the ramp-up stage. Therefore, the proposed knowledge-transfer method can facilitate knowledge inheritance from a mature diagnosis engine to the newer diagnosis engine, such that the new diagnosis engine can not only learn from its own failing boards with more intelligence, but also from other boards. To enable knowledge transfer, the cases from the previous diagnosis engine are first abstracted as descriptive knowledge. The descriptive knowledge is then interpreted in the context of the new diagnosis engine for learning. In practice, the sources of knowledge transfer can not only be several old boards that have similar functionality as the new board, but also a limited number of failing cases from the new board itself.

The knowledge-transfer method is based on topological analysis. To facilitate knowledge reuse, we divide the knowledge transfer into three stages: (i) hardware mapping, (ii) test program mapping, and (iii) case-knowledge transfer. The first two stages are reflected as root-cause mapping and syndrome mapping, respectively. The similarity between each component-pair of the old board and the new board is characterized by the functionality of the component on the board, while the similarity between test programs of the old and the new board is reflected by the topological characterization of the components covered by the test. Based on the root-cause mapping and syndrome mapping, the prior knowledge in the form of cases can be transferred to the new diagnosis system with a high degree of confidence.

We incorporate the synthetic cases generated from both knowledge discovery and knowledge transfer for training a diagnosis system. The weights of knowledge discovery and knowledge transfer are evaluated and discussed in detail in Sect. 7.5.

## 7.3 Knowledge-Discovery Method

Board-level diagnosis relies on syndromes recorded in a test log. The extraction of fault syndromes is critical for model training in a diagnosis system. The fault syndromes should provide complete description about the fails, and the extracted syndromes for different actions should have sufficient diversity such that we can eliminate ambiguity in the eventual repair recommendations. Fault syndromes vary across products and tests. For example, a segment of the log file for a failed traffic test is shown in Fig. 7.3. The fault syndromes extracted includes interrupt exception in SD12V_Master, mismatch errors in LAO Engine, and count error in SD23V, etc.

**Fig. 7.3** A segment of the
log file for a failed functional
test

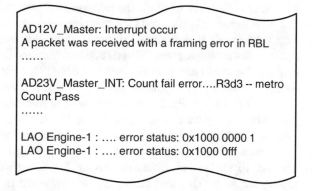

AD12V_Master: Interrupt occur
A packet was received with a framing error in RBL
......

AD23V_Master_INT: Count fail error....R3d3 -- metro
Count Pass
......

LAO Engine-1 : .... error status: 0x1000 0000 1
LAO Engine-1 : .... error status: 0x1000 0fff

Each of these elements is considered to be one syndrome. However, a traditional
diagnosis system treats each syndrome as a feature for classification and fails to
provide any information for understanding the failing mechanism.

The proposed knowledge-discovery method mines valuable information from syndromes, since the syndromes designed by the debug engineer also contain information
on underlying correlated root-causes. The proposed method follows a knowledge-
discovery procedure [10, 11]. The learning machine extracts the relationship between
test programs and components. This relationship is validated and converted to knowledge by test engineers. The discovered knowledge in turn helps in the update of the
learning machine.

From the perspective of test design at board level, each component on a board is
described or referred to using a knowledge pool of keywords. Each syndrome can
be treated as a piece of text, which contains many single words. When the words
in the syndrome match the keywords in the knowledge pool of a component, the
syndrome is more likely to cover the root-cause. In contrast, if the syndrome does
not contain any keyword related to the component, this syndrome is less likely to
have the capability of isolating the root-cause. Figure 7.4 illustrates the use of the

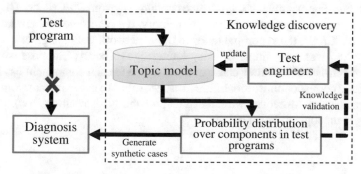

**Fig. 7.4** Flowchart of knowledge-discovery method

knowledge-discovery method. The relationship between root-causes and syndromes can be automatically revealed by a machine-learning technique called topic model. This relationship is represented in the form of a probability distribution involving components in test programs. Given such a probability distribution, test engineers can evaluate the relationship on the basis of which diagnosis system is trained. Moreover, test engineers can update the keyword list in the topic model based on the obtained probability distribution. Compared to rule-based or model-based methods that require significant manual effort from test engineers, the knowledge-discovery method significantly reduces the time needed to develop the diagnosis engine.

In machine learning or natural language processing, a *topic model* can discover the abstract "topics" that occur in a document [12]. Given a document about a particular topic, specific words about the topic tends to appear in the document with varying frequency. A document typically contains several topics in different proportions; thus the content on each topic may vary. A topic model can capture the proportion (a.k.a. probability) of topics in a document based on the statistics of words in the document and in the model for each topic. Our goal is to obtain $p(z_j)$, i.e., the probability that a document belongs to a topic $z_j$. Suppose that a document is regarded as a bag of words. The probability of each word $w_i$ in the document is $p(w_i) = \sum_{j=1}^{T} p(w_i|z_j)p(z_j)$, where $p(w_i|z_j)$ is the probability of word $w_i$ given the topic $z_j$ and $T$ is the total number of topics. Suppose there are a total of $N$ different words in $M$ documents, we include the probabilities of all words in an $N \times M$ matrix $P(w)$, which can be calculated as $P(w) = P(w|z) \times P(z)$, where $P(w|z)$ is a $N \times T$ matrix that corresponds to the probability distribution over words in each topic, and $P(z)$ is a $T \times M$ matrix corresponding to the probability distribution over topics in these documents. Therefore, P(z) can be calculated by

$$P(z) = \Theta \times P(w), \tag{7.1}$$

where $\Theta = (P(w|z)^T P(w|z))^{-1} P(w|z)^T$. To apply such a topic model to board-level diagnosis, root-causes can be regarded as topics, while syndromes can be regarded as documents. The more keywords related to a root-cause occur in the syndrome description, the more likely is it that the root-cause can be isolated by the syndrome. For example, if a syndrome contains keywords "ECC" or "DDR", the syndrome tends to target a certain type of memory. If a syndrome contains keywords "packet" or "CRC", the syndrome tends to target network traffic engine.

Figure 7.5 gives an example of using knowledge discovery. Suppose we have a board that consists of four root-cause components, and each component has its keyword list and corresponding probabilities, respectively. We can extract the knowledge as a matrix $P(w|z)$. Moreover, we can extract probability distribution $P(w)$ over keywords from the log, as shown below:

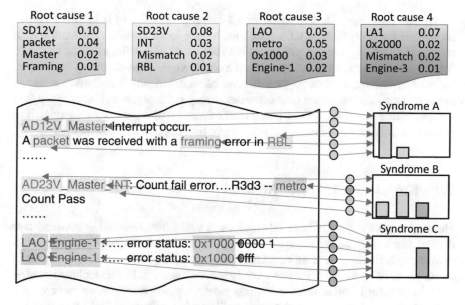

**Fig. 7.5**  A segment of the log file for a failed functional test

$$P(w) = \begin{bmatrix} 0.1 & 0 & 0 \\ 0.1 & 0 & 0 \\ 0.1 & 0.1 & 0 \\ 0.1 & 0 & 0 \\ 0 & 0.1 & 0 \\ 0 & 0.1 & 0 \\ 0 & 0 & 0 \\ 0.1 & 0 & 0 \\ 0 & 0 & 0.2 \\ 0 & 0.1 & 0 \\ 0 & 0 & 0.2 \\ 0 & 0 & 0.2 \\ 0 & 0 & 0 \\ 0 & 0 & 0 \\ 0 & 0 & 0 \end{bmatrix} \quad P(w|z) = \begin{bmatrix} 0.1 & 0 & 0 & 0 \\ 0.4 & 0 & 0 & 0 \\ 0.2 & 0 & 0 & 0 \\ 0.1 & 0 & 0 & 0 \\ 0 & 0.8 & 0 & 0 \\ 0 & 0.3 & 0 & 0 \\ 0 & 0.2 & 0 & 0.2 \\ 0 & 0.1 & 0 & 0 \\ 0 & 0 & 0.5 & 0 \\ 0 & 0 & 0.5 & 0 \\ 0 & 0 & 0.3 & 0 \\ 0 & 0 & 0.2 & 0 \\ 0 & 0 & 0 & 0.7 \\ 0 & 0 & 0 & 0.2 \\ 0 & 0 & 0 & 0.1 \end{bmatrix} \tag{7.2}$$

where each row corresponds to a word. By using Eq. (7.1), we can get $P(z)$ as below:

$$P(z) = \begin{bmatrix} 0.36 & 0.09 & 0 \\ 0.08 & 0.33 & 0 \\ 0 & 0.13 & 0.52 \\ 0 & 0 & 0 \end{bmatrix}. \tag{7.3}$$

We can thus conclude that syndrome 1 has a probability of 0.36 correlated to root-cause 1, a probability of 0.08 correlated to root-cause 2, and no correlation with root-cause 3 and 4. Similar syndrome and root-cause correlations can be obtained for the other two syndromes. By using the above knowledge-discovery technique, for each syndrome, a set of root causes can be automatically identified to relate to the syndrome. The proposed method considerably reduces manual labor when it is applied to rule-based and model-based methods, and it also provides an approach to collect knowledge that cannot be obtained in a traditional reasoning method.

## 7.4  Knowledge-Transfer Method

Human learners appear to have inherent ways to transfer knowledge between two similar tasks. That is, we recognize and apply relevant knowledge from previous learning experiences when we encounter new tasks. The more related a new task is to our previous experience, the more easily we can master it. In the machine learning domain, transfer learning is a challenging research problem; while we can enhance our knowledge of a task through transfer of knowledge from a related task that has already been learned, most machine learning algorithms are designed to address only a single task [13].

In previous work on reasoning-based diagnosis, classical methods such as artificial neural networks, support-vector machines, and decision trees [3, 4] were used for training a diagnosis system for a board. Diagnosis systems for different boards, even for a board with only slighted updated test programs, must be trained separately. Although an automated training mechanism facilitates the development of a diagnosis system, the data acquisition process for a diagnosis system can be tedious and impractical in an industry setting. During the initial product ramp-up phase, reasoning-based diagnosis is need for yield learning, but the required database is not available due to lack of volume. Therefore, an intelligent diagnosis system, which can not only "learn" the knowledge from itself, but also from other diagnosis systems, can offer the important benefit of more effective knowledge acquisition at lower cost.

The knowledge-transfer process can be divided into three stages: (1) root-cause mapping, (2) syndrome mapping, and (3) case-knowledge transfer. This partitioning (see Fig. 7.6) can be viewed as a realistic system functional-test design procedure as described below.

• Stage I mimics the board functionality design stage, where design engineers modify (i.e., add, remove, or change) the onboard components to get a new board design in order to meet design specifications. In contrast, since the design-for-test team has not yet joined the design process at this stage, the test program remains the same as for the old board. In Stage I, only the board structure is transformed, while the test programs are simply inherited. In practice, we need to label the component according to its functionality to distinguish one component from another. For example, a component is marked as boot ROM when it is in charge of initial system start-up. Another component is marked as CPU external memory when it is a ROM or RAM device

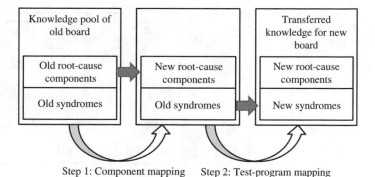

Step 1: Component mapping     Step 2: Test-program mapping

**Fig. 7.6** Two-stage knowledge transfer from an old diagnosis system to a new diagnosis system

connected to a CPU. Although components may vary from generation to generation or from board to board, those components with the same functionality label are prone to suffer similar functional failures. Therefore, we can reconstruct the board structure conceptually by leveraging functionality labels for components. For example, in Figs. 7.7 and 7.8, a component is labeled as ASIC B in Fig. 7.7. A similar component labeled as ASIC Bn can be found in Fig. 7.8. Although two components on the two board differ from each other in term of the manufacturer or provider, operating frequency, or number of pins, these two components play the same role on the board. We therefore conclude that ASIC B can be mapped to ASIC Bn, and thus the knowledge about ASIC B can be mapped to ASIC Bn.

In knowledge transfer, there are four types of relationships between components on the old board and those on the new board,

1. Suppose a component on the new board has exactly the same functionality label as on the old board. We can then map the component from the old board to the new board. Examples include ASIC B in Fig. 7.7 versus ASIC Bn in Fig. 7.8, and also Memory F in Fig. 7.7 versus Memory Fn and Hn in Fig. 7.8.
2. If the functionality provided by a component on the new board was not implemented on the old board, and is therefore reflected as a new functionality label

**Fig. 7.7** Illustration of a mature board from which knowledge can be transferred

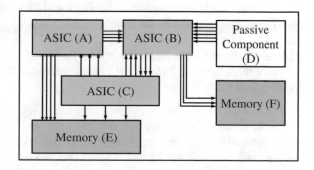

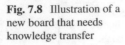

**Fig. 7.8** Illustration of a
new board that needs
knowledge transfer

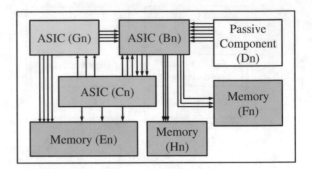

on a new board, we regard it as a new type of a root-cause. Clearly, there is no
knowledge of such type of a root-cause on the old board that can be transferred
to the new board. For example, in Fig. 7.8, ASIC Gn is a new component on the
new board, and it differs from any component on the old board.

3. A component on the old board is removed and the corresponding functionality of
   this component is not supported on the new board. The knowledge with regard
   to this component can no longer be transferred to the new board. For example in
   Fig. 7.7, ASIC A is a component on the old board that no longer exists on the new
   board.

4. A component on the new board has only partial functionality of the component
   on the old board, or the functionality of a component on the new board is a
   combination of several components on the old board. The functionality of the
   component on the new board is different from those on the old board. Therefore,
   the knowledge of this type of component cannot be transferred.

• Stage II mimics the test-program development stage. In this stage, the functionality has already met the board-design specifications. In order to assist design for
test/diagnosis, the test design team joins the design process to develop new test program for the new board structure. We focus here on the changes to the test programs,
while the board structure is considered to be fixed. Given the condition that two
boards are similar in terms of functionality, knowledge about most of the test programs for the old board can still be used for those for the new board. Consider the
examples shown in Figs. 7.9 and 7.10. ASIC B and Memory F on the old board in
Fig. 7.9 have similar components—ASIC Bn and Memory Fn—on the new board in
Fig. 7.10. Therefore, test $T_4$ for the old board can be viewed as being similar test $T_{4n}$
for the new board. We conclude therefore that $T_4$ can be mapped to $T_{4n}$, thus the
knowledge about $T_4$ can be transferred to $T_{4n}$.

In addition, there are three types of transformations for test-program development,
as described below:

1. If a test for the new board covers components with the same set of functionality
   labels as on the old board, we conclude that the knowledge of this test can be
   mapped to the new test program. If a case from the old board refers to a fail with
   the syndrome collected from this test, we can easily find a similar test for the

**Fig. 7.9** Illustration of tests on a mature board with knowledge to transfer from

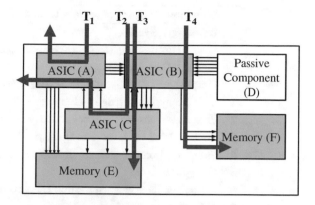

**Fig. 7.10** Illustration of tests on a new board that needs knowledge transfer

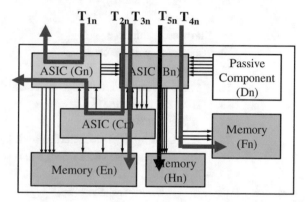

new board. For example, $T_4$ for the old board in Fig. 7.9 and $T_{4n}$ for new board in Fig. 7.10 correspond to such a pair. Also, $T_3$ can be mapped to $T_{3n}$ since both tests covers two ASICs and one memory chip of the same type. Note that $T_{5n}$ for the new board in Fig. 7.10 can also be mapped to $T_4$, because Fn and Hn can both be labeled with Memory and they are connected to Bn, resulting in both devices sharing the same topological characteristics.

2. If a test $T_i$ is designed exclusively for the new board and targets new components, we regard this test as being new. No knowledge about the tests for the old board can be transferred to $T_i$ in the new test program. For example, in Fig. 7.10, $T_{2n}$ and $T_{1n}$ are both new because the component Gn covered by $T_{2n}$ and $T_{1n}$ is new.

3. If a test $T_i$ covers components that exists only on the old board, knowledge about $T_i$ must not be transferred to the new board to avoid biased or misleading knowledge. Examples include $t_2$ and $t_1$, as shown in Fig. 7.9.

• Stage III leverages the root-cause mapping and syndrome-mapping information obtained from Stage I and II. The root-cause and syndromes for a case can be abstracted in terms of functionality labels at a higher knowledge level. Thus, the abstract knowledge can be interpreted in the context of the new board to generate synthetic cases with high confidence.

---

**Algorithm 1:** Knowledge-transfer method

---

**Input:** $<rc, \textbf{Synd}>$,

      $\textbf{Dict}_{old} <rc_{old}, l_{old}>$, $\textbf{Dict}_{new} <rc_{new}, l_{new}>$

      $\textbf{RC}_{old}$, $\textbf{L}_{old}$, $\textbf{RC}_{new}$, $\textbf{L}_{new}$

**Output:** $\textbf{S} : \{<rc', \textbf{Synd}'> | rc' \in \textbf{RC}_{new}, \textbf{Synd}' \subseteq \textbf{Synd}_{new}\}$

1   $\textbf{S} \leftarrow$ Empty, **Knowledge** $\leftarrow$ Empty;

2   //Map root-cause to abstract knowledge (label)

3   Look up label $l \in \textbf{L}$ for $rc$ in $\textbf{Dict}_{old}$;

4   //Map Syndrome to abstract knowledge (a set of labels)

5   **For** $synd$ in **Synd**

6      Look up label characterization $\textbf{L}$ for $synd$;

7      **Knowledge**.add($\textbf{L}$);

8   **End for**

9   **Knowledge**.unique();

10  //Construct synthetic cases

11  **For** $rc' \in \textbf{RC}_{new}$

12     //Map root-cause knowledge to actually root-causes

13     **If** $rc'$ has label $l$ in $\textbf{Dict}_{new}$

14       **For** $\textbf{L} \in$ **Knowledge**

15         Look up label characterization $\textbf{L}_{new}$ for $\textbf{Synd}'$;

16         //Map syndrome knowledge to actual syndromes

17         **If** $\textbf{L}_{new} = \textbf{L}$

18           $\textbf{S}$.add($<rc', \textbf{Synd}'>$);

19         **End if**

20       **End for**

21     **End if**

22  **End for**

23  **return** $\textbf{S}$;

---

**Fig. 7.11** Proposed knowledge-transfer algorithm

Figure 7.11 describes the knowledge-transfer algorithm. We are given a real case $<rc, \textbf{Synd}>$ collected from the old board, where $rc$ is the root-cause and **Synd** refers to a set of syndromes for the failing board. Our goal is to generate a set of synthetic cases $\textbf{S} : \{<rc', \textbf{Synd}'> | rc' \in \textbf{RC}_{new}, \textbf{Synd}' \subseteq \textbf{Synd}_{new}\}$ for the new board, where $\textbf{RC}_{new}$ is the root-cause set and $\textbf{Synd}_{new}$ is the syndrome set for the new board, respectively. We have two dictionaries that consist of the components and the corresponding labels, i.e., $\textbf{Dict}_{old} <rc_{old}, l_{old}>$ for the old board and $\textbf{Dict}_{new} <rc_{new}, l_{new}>$ for the new board, respectively, where each root-cause is paired to its corresponding functionality label.

First, we carry out root-cause mapping. The functionality label of root-cause $rc$ is identified to be $l$ by looking up $\mathbf{Dict}_{old}$. Similarly, by looking up functionality label $l$ in $\mathbf{Dict}_{new}$ for the new board, a set of root-causes $\{rc_{new} \mid < rc_{new}, l > \in \mathbf{Dict}_{new}\}$ can be found with the same functionality label $l$. Next, we proceed with syndrome mapping. Each syndrome $synd \in \mathbf{Synd}$ from the case can be mapped to $synd_{new}$ for the new board based on label characterization. A label-characterization procedure refers to description of a syndrome in the form of a set of functionality labels $\mathbf{L}$. We need to find a set of syndromes $\mathbf{Synd}'$ for the new board with the same label characterizations as that of $\mathbf{Synd}$ for the old board. Due to the fact that multiple components on the new board can have the same functionality label $l$, multiple syndromes $synd_{new}$ may have the same label characterization $\mathbf{L}_{new}$. A set of synthetic cases $\mathbf{S}$ can be generated for the new board by assuming that root-cause $r_{new}$ has the same functionality label $l$ as that of the root-cause on the old board. Moreover, the corresponding syndromes for the synthetic case have the same label characterization as that of the syndromes for the old board. Assuming that the number of syndromes for an actual case is much less than the total number of syndromes, the computation complexity of the algorithm in Fig. 7.11 is O(UV), where U is the total number of syndromes and V is the total number of root-causes for the new board, respectively.

We illustrate the knowledge-transfer method with an example. Suppose that we have a failing board from the old board with the structure shown in Fig. 7.7 and the test program in Fig. 7.9. A failing case has a root-cause—memory F, and the corresponding syndrome is a "fail" for $T_4$. Thus, the case to be transferred is $< F, T_4 >$. First, we use root-cause mapping. The components Fn and Hn on the new board have the same functionality label "Memory" as F on the old board. Next, we carry out syndrome mapping. We find that tests $T_{4n}$ and $T_{5n}$ have the same functionality characterization as $T_4$, because they both consist of a Memory component and a ASIC B component. Therefore, we can generate two synthetic cases $< Fn, t_{4n} >$ and $< Hn, T_{5n} >$, respectively during the knowledge-transfer step.

## 7.5   Results

Experiments were performed on five industrial boards, two of which are currently in high-volume production and the other three of which are in the ramp-up stage. Relevant information about the boards is provided in Table 7.1 and Table 7.2. For example, Board $1'$ in Table 7.1 has a mature diagnosis engine that has been

**Table 7.1** Information about the industrial boards with mature diagnosis engines

|  | Board $1'$ | Board $2'$ |
| --- | --- | --- |
| Number of syndromes | 299 | 201 |
| Number of repair candidates (components) | 153 | 116 |
| Number of boards | 1903 | 1634 |

**Table 7.2** Information about the industrial boards at the ramp-up stage

|                                          | Board 1 | Board 2 | Board 3 |
|------------------------------------------|---------|---------|---------|
| Number of syndromes                      | 3474    | 5537    | 10884   |
| Number of repair candidates (components) | 258     | 276     | 483     |
| Number of boards                         | 274     | 193     | 245     |

maintained for nearly two year. A total of 1903 repaired boards were collected from the contract manufacturer's database. A total of 299 fault syndromes were extracted from failure logs. The number of faulty components identified in the database for repair action is 153. Board 1 and Board 2 in Table 7.2 are the boards at the ramp-up stage with only a few failing boards. These boards share similar designs, where Board 1, 2, and 3 are the successor designs to Board $1'$ and $2'$. The design changes include increasing/decreasing network bandwidth, cost-driven board redesign with component changes, etc. Therefore, knowledge-transfer method can be applied to these boards.

The knowledge-discovery and knowledge-transfer algorithms are implemented using Python. We use several diagnosis engines for demonstration, including support-vector machines(SVMs), artificial neural networks(ANNs), and decision trees(DTs) based on weka [14]. Experiments were run on a 64-bit Linux system with 12 GB of RAM and a quadcore Intel i7 processor running at 2.67 GHz.

In order to assess the performance of the knowledge-discovery and transfer methods, and its ability to accurately predict the root-cause of failure on a new board, we use the real cases collected for three new boards described in Table 7.2 to evaluate the diagnosis engines. To ensure real-time diagnosis and repair, we assume that we are allowed at-most three attempts to replace the potential failing components. Success ratio (SR) is the ratio of the number of correctly diagnosed cases to the total number of cases. For example, if ten failing cases are diagnosed, a SR of 70 % means that seven out of ten cases are correctly diagnosed and the remaining three cases are incorrectly diagnosed. We define $SR_1$ as the success ratio corresponding to the cases for which the board is deemed to have been successfully repaired only when the actual faulty component is identified and placed at the top of the list of root-cause candidates. We also define $SR_2$ ($SR_3$) as the success ratio corresponding to the case that a board is deemed to have been successfully repaired if the actual faulty component is in the first two (three) positions in the list of candidates.

In Figs. 7.12, 7.13 and 7.14, we evaluate the diagnosis results for Boards 1–3 by comparing the proposed knowledge-discovery and transfer framework to the original knowledge pool created manually by the test designers; we refer to the latter as the "Manual method". The creation of such a comprehensive knowledge pool takes considerable time—a week or more for a team of engineers working on a board design—and it requires several experienced technicians to maintain the knowledge pool for months or more. "Knowledge discovery" uses the proposed knowledge-discovery method only (i.e., without knowledge transfer). "Knowledge transfer$_{b=1}$"

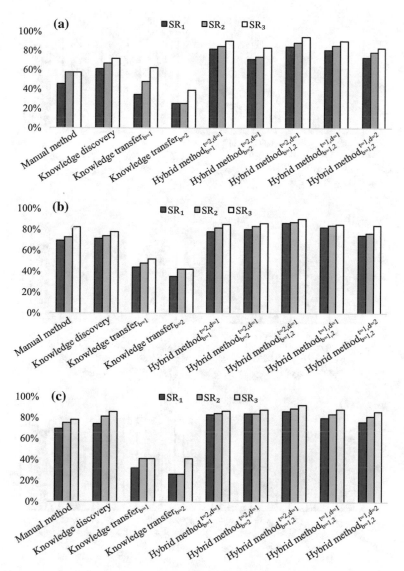

**Fig. 7.12** Comparison of diagnosis accuracy using SVMs between manual method, knowledge discovery, knowledge transfer, and the hybrid method. **a** Board 1. **b** Board 2. **c** Board 3

uses the proposed knowledge transfer (i.e., without knowledge discovery) based on the knowledge from the mature diagnosis system for Board $1'$, and "Knowledge transfer$_{b=2}$" uses that from Board $2'$ only. The hybrid method combines knowledge discovery and knowledge transfer by using synthetic cases generated from both sources as shown in Fig. 7.2; "Hybrid method$_{b=1}$" learns knowledge from Board $1'$ only; "Hybrid method$_{b=2}$" acquires knowledge from Board $2'$ only; and "Hybrid

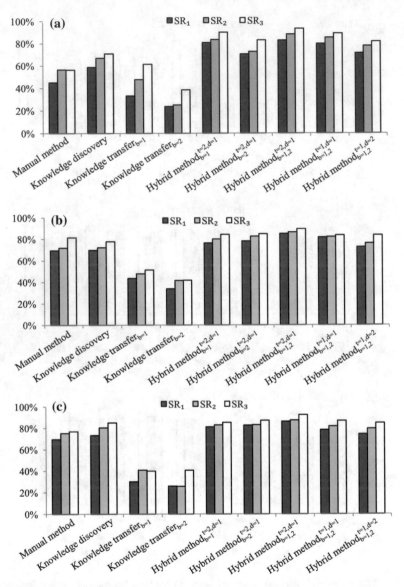

**Fig. 7.13** Comparison of diagnosis accuracy using ANNs between manual method, knowledge discovery, knowledge transfer, and the hybrid method. **a** Board 1. **b** Board 2. **c** Board3

method$_{b=1,2}$" acquires knowledge from both Board 1′ and Board 2′. Moreover, we evaluate hybrid methods with different weight configurations for knowledge discovery and knowledge transfer. For example, "Hybrid method$^{t=1,d=2}$" means the weight of synthetic cases generated using knowledge transfer is half of that using knowledge discovery; "Hybrid method$^{t=1,d=1}$" means the weights of synthetic cases

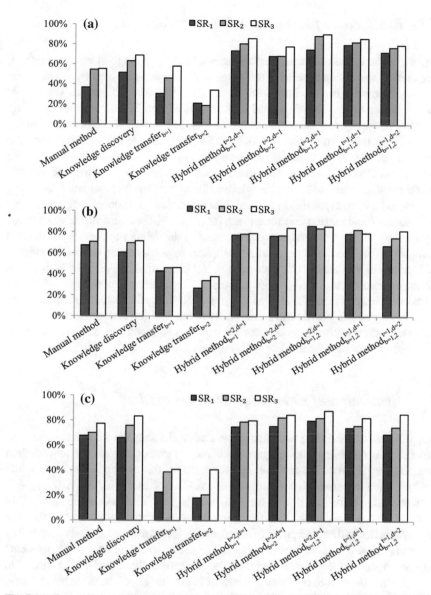

**Fig. 7.14** Comparison of diagnosis accuracy using DTs between manual method, knowledge discovery, knowledge transfer, and the hybrid method. **a** Board 1. **b** Board 2. **c** Board 3

generated using knowledge transfer and knowledge discovery are same; and "Hybrid method$^{t=2,d=1}$" means the weight of synthetic cases generated using knowledge discovery is twice of that using knowledge discovery.

### 7.5.1  Evaluation of Knowledge-Discover Method

First, we evaluate the diagnosis performance of the knowledge-discovery method. Compared to the manual method, the diagnosis accuracy obtained by knowledge-discovery method is higher. For example, for the SVM-based diagnosis system for Board 1 in Fig. 7.12a, $SR_1$, $SR_2$, and $SR_3$ are 61, 67 and 72 %, respectively, using the knowledge-discovery method, compared to 45, 57 and 57 %, respectively, using the manual method. An increase of 16 % in $SR_1$ is therefore obtained by using the knowledge-discovery method. Increases in success ratios are also observed in both Board 2 and Board 3. The reason that the knowledge-discovery method outperforms the manual method is that knowledge discovery avoids errors and biases that are likely to be introduced due to manual data input. Although debug engineers have many years of hands-on experience and are familiar with the board structure and test-program designs, the knowledge pool generated by the debug engineers is biased and incomplete, which can lead to low diagnosis accuracy. Moreover, due to the significant increase in the number of syndromes in test programs, it takes prohibitive long time to generate the knowledge pool manually. Nevertheless, knowledge-discovery method still suffers from the loss in diagnosis accuracy in that, in practice, error scenarios vary across boards and sometimes are not straightforward from the observation on syndromes. Such cases occurs in volume manufacturing and they can only be learned in reasoning-based diagnosis engines [5, 6].

### 7.5.2  Evaluation of Knowledge-Transfer Method

Next, we evaluate the diagnosis accuracy achieved using the knowledge-transfer method. We observe that the diagnosis accuracy is comparable to manual method when we use "Knowledge-transfer$_{b=2}$" for training the diagnosis engines, while it is low when we use "Knowledge-transfer$_{b=1}$". For example, the $SR_1$ is 34 % for "Knowledge-transfer$_{b=1}$" and the $SR_1$ is 25 % for "Knowledge-transfer$_{b=2}$" in Fig. 7.12a. The reason for the difference in diagnosis accuracies obtained using the two methods lies in the differences in board structure and test-program design. When we compare Board 1′ to the three new boards, we find that a number of components on the new boards have different functionality labels from those on the old boards. On the other hand, the design of Board 2′ is closer to the designs of the three new boards, which makes the knowledge-transfer method more effective. Compared to the knowledge discovery and the manual method, knowledge transfer exploits the knowledge that was accumulated for the other boards in volume production. This knowledge can be transferred with high confidence, but the knowledge-transfer method is also limited by functional dissimilarities between the knowledge-source board and new board (e.g., it is impossible and meaningless to transfer the diagnosis knowledge from a linear-process unit to an embedded microcontroller.) Nevertheless diagnosis knowledge derived using the knowledge-transfer method is still valuable for training the diagnosis engine of new boards.

### 7.5.3   Evaluation of Hybrid Method

Then, we evaluate the hybrid method. First, we compare the hybrid methods with knowledge of different previous boards. For example, in the case of Board 1 in Fig. 7.12a, $SR_1$, $SR_2$, and $SR_3$ are 84.5, 88.5 and 94.5 %, respectively, for "Hybrid method$_{b=1,2}^{t=2,d=1}$". Compared to the manual method, an increase of 39 % in $SR_1$ is obtained by using "Hybrid method$_{b=1,2}^{t=2,d=1}$". The large increase in success ratios in $SR_1$, $SR_2$, and $SR_3$ can also be observed for Board 2 in Fig. 7.12b and Board 3 in Fig. 7.12c, respectively. The reason for this remarkable increase in diagnosis accuracy is that the hybrid method leverages the benefits of both knowledge discovery and knowledge transfer; the knowledge-discovery method captures the knowledge of relationship between syndromes and root-causes, while knowledge-transfer method utilizes knowledge from old boards that exhibit similar failures as the new board.

Next, we compare the hybrid method using different weight configurations. We observe that the diagnosis accuracy is higher when we assign higher weight to the knowledge-transfer method. For example, in the case of Board 1 in Fig. 7.12a, the $SR_1$ are 83, 81 and 74 %, respectively, for "Hybrid method$_{b=1,2}^{t=2,d=1}$", "Hybrid method$_{b=1,2}^{t=1,d=2}$", "Hybrid method$_{b=1,2}^{t=1,d=2}$", respectively. The reason for this observation is that knowledge transfer brings knowledge that has been successfully utilized for the earlier-generation boards with similar design. This knowledge reveals the inherent relationship between syndromes and root-causes, which cannot be obtained easily using the knowledge-discovery method or the manual method. The use of higher weight for knowledge transfer can help correct the diagnosis results.

In addition, we compare the diagnosis systems using statistical evaluation metrics, namely *precision* and *recall*, described in Chap. 5.

Figures 7.15 and 7.16 show the precision and recall distributions for all the root-causes for the diagnosis system trained using different methods for Board 1. In each graph, the x-axis corresponds to the value range of *precision* or *recall*, while the y-axis corresponds to the number of root-causes that have the precision/recall value within the corresponding value range. A total of 258 root causes are available for detailed examination for Board 1. We observe that precision and recall values of root-causes for the hybrid methods are higher than the manual method. For example, the recall values of 188 root causes are 1 and the precision values of 179 root causes are 1 for "Hybrid method$_{b=1,2}^{t=2,d=1}$"; in contrast, the recall values of only 145 root causes are 1 and the precision values of 170 root causes are 1 for the manual method. Moreover, these values are considerably lower for knowledge transfer alone. For example, the precision values of 109 root causes are below 0.1, and the recall values of 89 are below 0.1 for Knowledge transfer$_{b=1}$. The reason for low precision and recall values can be attributed to the fact that most components on Board 1' have different functionalities compared to Board 1. No knowledge can be transferred if components of similar functionalities cannot be found on the new board.

We also compare the diagnosis accuracy obtained using difference types of diagnosis engines. We observed that the difference in diagnosis accuracy is not significant. The diagnosis performance of SVM-based diagnosis system outperforms that

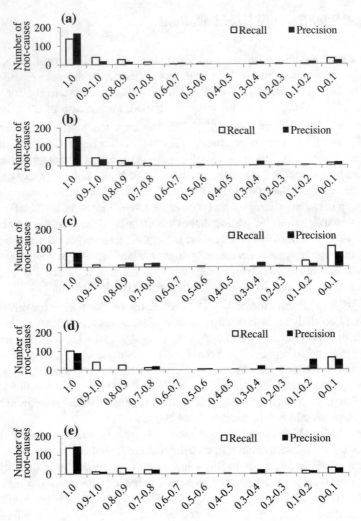

**Fig. 7.15** Precision and recall distribution for an SVM-based diagnosis system for Board 1 (Part I). **a** Manual Method. **b** Knowledge discovery. **c** Knowledge transfer$_{b=1}$. **d** Knowledge transfer$_{b=2}$. **e** Hybrid method$_{b=1}^{t=2,d=1}$

of ANN-based and DT-based diagnosis systems as illustrated in the literatures [3, 4]. As shown in Figs. 7.12, 7.13 and 7.14, knowledge discovery and knowledge transfer are both effective for different types of diagnosis engines.

In addition, the knowledge-discovery method can also be useful in practice as a reference for debug engineers to evaluate the knowledge pool generated by the manual method. The proposed approach based on knowledge discovery and knowledge transfer is recommended for a reasoning-based diagnosis engine since it facilitates automation in diagnosis, therefore overcoming the bottleneck of "knowledge acquisition" at the product ramp-up stage (Fig. 7.16).

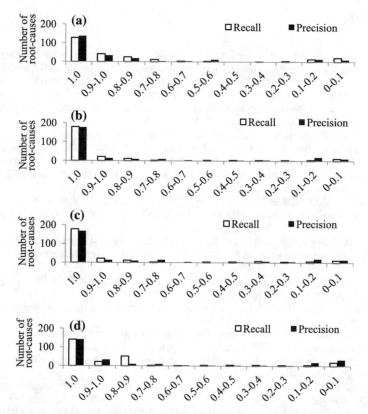

**Fig. 7.16** Precision and recall distribution for an SVM-based diagnosis system for Board 1 (Part II). **a** Hybrid method$_{b=2}^{t=2,d=1}$. **b** Hybrid method$_{b=1,2}^{t=2,d=1}$. **c** Hybrid method$_{b=1,2}^{t=1,d=1}$. **d** Hybrid method$_{b=1,2}^{t=2,d=2}$

## 7.6 Chapter Summary

Reasoning-based diagnosis method suffers low diagnosis accuracy due to the difficulty of "data acquisition" at the product ramp-up stage. This chapter has proposed a knowledge-discovery and knowledge-transfer framework to automatically build a knowledge pool for the new diagnosis engine from the information contained in the syndromes as well as the diagnosis experience collected from other diagnosis engines of similar board types. Five industrial boards have been used to demonstrate the effectiveness of using the knowledge-discovery and knowledge-transfer framework for root-cause identification in failing boards.

# References

1.  Fenton W, McGinnity T, Maguire L (2001) Fault diagnosis of electronic systems using intelligent techniques: a review. IEEE Trans Syst Man Cybern Part C: Appl Rev 31:269–281
2.  Watson I, Marir F (1994) Case-based reasoning: a review. Knowl Eng Rev 9(4):327–354
3.  Ye F, Zhang Z, Chakrabarty K, Gu X (2013) Board-level functional fault diagnosis using artificial neural networks, support-vector machines, and weighted-majority voting. IEEE Trans Comput-Aided Des Int Circuits Syst (TCAD) 32(5):723–736
4.  Ye F, Zhang Z, Chakrabarty K, Gu X (2012) "Adaptive board-level functional fault diagnosis using decision trees". In: Proceedings of the IEEE Asian test symposium (ATS), pp. 202–207
5.  Ye F, Zhang Z, Chakrabarty K, Gu X (2012) "Board-level functional fault diagnosis using learning based on incremental support-vector machines". In: Proceedings of the IEEE Asian test symposium (ATS), pp. 208–213
6.  Bolchini C, Quintarelli E, Salice F, Garza P (2013) "A data mining approach to incremental adaptive functional diagnosis", In: Proceedings of the IEEE international symposium on defect and fault tolerance in VLSI systems (DFT), pp. 13–18
7.  Eklow B, Hossein A, Khuong C, Pullela S, Vo T, Chau H (2004) "Simulation based system level fault insertion using co-verification tools". In: Proceedings of the IEEE international test conference (ITC), pp. 704–710
8.  Manley D, Eklow B, (2002) "A model based automated debug process". In: Proceedings of the IEEE board test workshop, pp. 1–7
9.  Feret MP, Glasgow JI (1997) Combining case-based and model-based reasoning for the diagnosis of complex devices. Appl Intell 7(1):57–78
10. Wang L.-C (2013) "Data mining in design and test processes: basic principles and promises". In: Proceedings of the IEEE international symposium on physical design, pp. 41–42
11. Klösgen W, Zytkow JM (2002) Handbook of data mining and knowledge discovery. Oxford University Press, Oxford
12. Steyvers M, Griffiths T (2007) Probabilistic topic models. Handb latent semant anal 427(7):424–440
13. Pan S, Yang Q (2010) A survey on transfer learning. IEEE Trans Knowl Data Eng 22(10):1345–1359
14. Hall M, Frank E, Holmes G, Pfahringer B, Reutemann P, Witten IH (2009) The weka data mining software: an update. ACM SIGKDD Explor Newsl 11(1):10–18

# Chapter 8
# Conclusions

Advances in semiconductor technology have driven the electronics industry toward higher levels of integration with shorter development cycles. However, ever-increasing integration density and clock frequencies make it more difficult to ensure product quality at the board and system levels, which adversely impacts the competitiveness of a system manufacturer. Automated test and diagnosis systems are thus needed to ensure high quality and manage production cost. However, existing board-level testing and diagnostic strategies are insufficient to meet the requirements of the user community for products with high reliability. The development and maintenance of diagnosis systems using existing solutions require extensive human experience and manual labor. Moreover, today's diagnosis systems suffer low accuracy, which leads to an increase in the number of repair attempts and low product yield. The resulting high manufacturing cost motivates the need for an automated diagnosis system with high diagnosis accuracy. This book has presented a number of solutions to address pressing challenges in board-level functional diagnosis, which forms then core of an automated diagnosis system.

This book has presented a set of intelligent diagnostic methods to reduce the dependence of board-level diagnosis on time-consuming and ineffective human effort. Multiple machine learning and statistical methods have been studied and adapted for diagnosis. Substantial improvement has been achieved over currently deployed diagnostic software. These solutions are not limited to a particular product; they are generic and can therefore be applied to various products. Although the goal of this book was to advance board-level diagnosis, the core techniques described in this book can also be leveraged for larger electronic systems.

The book covered technologies that significantly advanced board-level functional fault diagnosis. Recall that the diagnosis procedure involves several key components: (1) design of functional test programs, (2) collection of functional-failure syndromes, (3) building of the diagnosis engine, (4) isolation of root causes, and (5) evaluation of the diagnosis engine. This book has targeted all the above five components of the diagnosis procedure.

Chapter 2 proposed a diagnosis system that extends a support-vector machine-based diagnosis engine. The proposed diagnosis system provides high accuracy

© Springer International Publishing Switzerland 2017
F. Ye et al., *Knowledge-Driven Board Level Functional Fault Diagnosis*,
DOI 10.1007/978-3-319-40210-9_8

and incorporates self-learning based on multi-kernel support-vector machines (MK-SVMs). The MK-SVM method leverages a linear combination of single kernels to achieve accurate root cause isolation. Incremental learning is used to continuously adapt the diagnosis system to newly occurring error scenarios.

Chapter 3 described an enhanced diagnosis system that leverages data fusion based on majority-weighted voting (MWV). The MWV-based diagnosis system can leverage multiple learning techniques (e.g., ANNs, SVMs, BIs, etc.) for diagnosis. It offers an attractive approach for aggregating decisions obtained from multiple diagnosis engines and can thus achieve high diagnosis accuracy.

Chapter 4 proposed an adaptive diagnosis system based on decision trees. Due to the large number of syndromes required to ensure diagnosis accuracy, the diagnosis time can be considerable for complex boards. The proposed adaptive diagnosis method aims to reduce the diagnosis time based on statistical analysis using decision trees. The number of syndromes required for diagnosis can thus be significantly reduced compared to the number of syndromes used for system training. Moreover, in order to boost the diagnosis accuracy at an early stage in volume production and to bridge the knowledge obtained at test-design stage with the knowledge gained during volume production, the DT structure has been enhanced to enable incremental learning.

Chapter 6 targeted the preprocessing of syndromes for diagnosis. Traditional diagnosis systems fail to provide appropriate repair suggestions when the diagnostic logs are fragmented and some syndromes are not available. The proposed diagnosis system adds the capability to handle syndromes based on imputation methods. The use of imputation methods help us to infer missing syndromes, so as to aid in the isolation of root-cause components for those failing boards that cannot be diagnosed using traditional diagnosis engines.

Chapter 5 described an evaluation and enhancement framework for guiding diagnosis systems. The proposed evaluation framework leverages syndrome and root-cause analysis based on information theory. Syndrome analysis based on subset selection provides a representative set of syndromes with minimum redundancy and maximum relevance. Root-cause analysis measures the discriminative ability of differentiating a given root cause from others. The metrics obtained from the proposed framework can not only grade the performance of the diagnosis system, but also provide guidelines for test redesign to enhance diagnosis.

Finally, Chap. 7 presented two knowledge-acquisition methods to tackle the difficulty of "data acquisition" during the initial product ramp-up phase. The proposed knowledge-discovery method and the knowledge-transfer method help enrich the training data set for training diagnosis systems, thus overcoming the bottleneck of low diagnosis accuracy at an early stage during volume production. First, knowledge-discovery method based on nature-language processing is used to automatically mine the relationship between syndromes and components from syndrome descriptions. Second, given the effectiveness of syndromes for root-cause mapping, knowledge from diagnosis engines used for earlier-generation products can be automatically transferred for training diagnosis systems for new products.

## Directions for Future Work

Machine learning provides an important theoretical basis for advancing board- and system-level test and diagnosis, especially as we move toward more complex products with even higher integration levels and production in high volume. Inefficient diagnosis based on manual debug and repair will eventually be replaced by automated diagnosis and repair. Machine learning and computer-aided statistical analysis provide a promising solution for diagnosis automation. This book has opened up the following interesting new directions.

- **Multiple root-causes**

  A single root cause has been targeted for all failing boards in this book; in reality, however, complex systems often fail due to multiple failing components. In current diagnosis systems, multiple suspicious root causes are listed for repair by technicians with the most-likely root cause at the top of the list. However, syndromes obtained from failing systems can indicate more than one root cause. Error scenarios for multiple root causes cannot be handled using current diagnostic systems. It is important to develop a diagnosis system that can implicitly target multiple root causes. Such a system requires breakthroughs in training and diagnosis. In today's learning-based diagnosis systems, a learning case is formatted as a set of syndromes and a single root cause, but the desired multiple-root-cause diagnosis system must learn from experience in the form of a set of syndromes and multiple root causes.

- **Data mining in gate-level diagnosis**

  As machine learning techniques are increasingly adopted for board- and system-level diagnosis, we expect insights to emerge for utilizing machine learning for chip- or gate-level test and diagnosis. Current state-of-the-art test and diagnosis techniques at gate level mainly rely on structural tests, where it is assumed that we can preciously control chip voltages, temperatures, and signal levels. However, the process-variation profile varies for different chips, even if they are fabricated on the same wafer, and process variation is even more pronounced for different wafers. Different process variations lead to different error scenarios, which cannot be captured by structural tests. A data-driven reasoning-based system that utilizes machine learning can be a potential solution can be a solution for this problem.

  The massive amount of data collected from the same product that are manufactured earlier can be used to train a reasoning system, which can be used for future test or diagnosis. Moreover, given the ability of collecting monitoring data from the chip under operation, the reasoning system thus trained can be used to monitor or predict chip performance on-the-fly. The prediction system can thus also incorporate aging effects. In this research direction, hardware implementation of machine learning techniques is another big concern. Trade-off in hardware design (i.e., silicon area, response timing, prediction accuracy, test or diagnosis function) should be carefully evaluated.

- **Diagnosis automation**

  Although the diagnosis system studied in this book has involved considerable automation based on machine learning and artificial intelligence techniques, automation has not been fully realized for all the diagnosis steps. Research on automation for the following steps are needed:

  (1) Automatic identification of relationships between syndrome keywords to related components in knowledge discovery. A possible solution is to enable standard labeling for all components and analysis on board structural designs;

  (2) Automated improvement of tests based on evaluation framework. Current evaluation frameworks target the analysis of syndromes and root causes with weak diagnosis ability, while the test engineer must manually analyze the evaluation results to identify the appropriate test program for isolating the root cause with weak diagnosis ability from the remaining root causes. By analyzing the information of board designs, it is likely that appropriate test improvement can be automated.

# Index

© Springer International Publishing Switzerland 2017
F. Ye et al., *Knowledge-Driven Board-Level Functional Fault Diagnosis*,
DOI 10.1007/978-3-319-40210-9

Printed in the United States
By Bookmasters